# SpringerBriefs in Applied Sciences and Technology

T0171955

SpringerBriefs present concise summaries of cutting-edge research and practical applications across a wide spectrum of fields. Featuring compact volumes of 50 to 125 pages, the series covers a range of content from professional to academic.

Typical publications can be:

- A timely report of state-of-the art methods
- An introduction to or a manual for the application of mathematical or computer techniques
- A bridge between new research results, as published in journal articles
- A snapshot of a hot or emerging topic
- An in-depth case study
- A presentation of core concepts that students must understand in order to make independent contributions

SpringerBriefs are characterized by fast, global electronic dissemination, standard publishing contracts, standardized manuscript preparation and formatting guidelines, and expedited production schedules.

On the one hand, **SpringerBriefs in Applied Sciences and Technology** are devoted to the publication of fundamentals and applications within the different classical engineering disciplines as well as in interdisciplinary fields that recently emerged between these areas. On the other hand, as the boundary separating fundamental research and applied technology is more and more dissolving, this series is particularly open to trans-disciplinary topics between fundamental science and engineering.

Indexed by EI-Compendex, SCOPUS and Springerlink.

More information about this series at http://www.springer.com/series/8884

Sumit Bhowmik · Divya Zindani

# Hybrid Micro-Machining Processes

 Springer

Sumit Bhowmik
Department of Mechanical Engineering
National Institute of Technology Silchar
Silchar, Assam, India

Divya Zindani
Department of Mechanical Engineering
National Institute of Technology Silchar
Silchar, Assam, India

ISSN 2191-530X                          ISSN 2191-5318   (electronic)
SpringerBriefs in Applied Sciences and Technology
ISBN 978-3-030-13038-1              ISBN 978-3-030-13039-8   (eBook)
https://doi.org/10.1007/978-3-030-13039-8

Library of Congress Control Number: 2019931521

This Springer imprint is published by the registered company Springer Nature Switzerland AG
The registered company address is: Gewerbestrasse 11, 6330 Cham, Switzerland

# Contents

**1 Overview of Hybrid Micro-manufacturing Processes** . . . . . . . . . . . .  1
   1.1 Introduction . . . . . . . . . . . . . . . . . . . . . . . . . . . . . . . . . . . . . . . .  1
   1.2 Classification of Hybrid Micro-manufacturing Processes . . . . . . . .  2
       1.2.1 Compound Processes . . . . . . . . . . . . . . . . . . . . . . . . . . . . . .  3
       1.2.2 Energy-Assisted Micromachining Processes. . . . . . . . . . . . .  4
       1.2.3 Combined Hybrid Micromachining Processes . . . . . . . . . .  8
   1.3 Future Research Opportunities. . . . . . . . . . . . . . . . . . . . . . . . . . . .  8
   1.4 Conclusion. . . . . . . . . . . . . . . . . . . . . . . . . . . . . . . . . . . . . . . . . .  9
   References . . . . . . . . . . . . . . . . . . . . . . . . . . . . . . . . . . . . . . . . . . . .  10

**2 Laser-Assisted Micromachining** . . . . . . . . . . . . . . . . . . . . . . . . . . . .  13
   2.1 Introduction . . . . . . . . . . . . . . . . . . . . . . . . . . . . . . . . . . . . . . . . .  13
   2.2 Conceptual Framework to Laser Beam Micromachining
       (LBMM) . . . . . . . . . . . . . . . . . . . . . . . . . . . . . . . . . . . . . . . . . . . .  14
   2.3 Mechanism and Principle of Laser Ablation . . . . . . . . . . . . . . . . .  17
       2.3.1 Laser Ablation Process Through Nanosecond Laser . . . . . .  17
       2.3.2 Laser Ablation Process Through Picosecond Laser . . . . . .  18
       2.3.3 Laser Ablation Process Through Femtosecond Laser . . . . .  18
   2.4 Laser-Assisted Variant of Hybrid Micromachining Process . . . . . .  20
       2.4.1 Laser Beam Assisted Water Jet Micromachining . . . . . . . .  20
       2.4.2 Laser Beam Assisted Jet Electrochemical
             Micromachining . . . . . . . . . . . . . . . . . . . . . . . . . . . . . . . . .  21
       2.4.3 Laser Beam Assisted Micro-milling/Grinding Process . . . .  22
   2.5 Conclusion. . . . . . . . . . . . . . . . . . . . . . . . . . . . . . . . . . . . . . . . . .  22
   References . . . . . . . . . . . . . . . . . . . . . . . . . . . . . . . . . . . . . . . . . . . .  23

**3 Magnetic Field Assisted Micro-EDM** . . . . . . . . . . . . . . . . . . . . . . .  25
   3.1 Introduction . . . . . . . . . . . . . . . . . . . . . . . . . . . . . . . . . . . . . . . . .  25
   3.2 Electrical Discharge Machining (EDM) and Micro-EDM
       Process . . . . . . . . . . . . . . . . . . . . . . . . . . . . . . . . . . . . . . . . . . . . .  26

3.3 Micro-EDM Process and Its Mechanics . . . . . . . . . . . . . . . . . . . . 29
   3.3.1 Mechanism of Material Removal . . . . . . . . . . . . . . . . . . . . 29
   3.3.2 Process Parameters . . . . . . . . . . . . . . . . . . . . . . . . . . . . . . 30
3.4 Working Mechanism of Magnetic Field Assisted Micro-EDM
   Process . . . . . . . . . . . . . . . . . . . . . . . . . . . . . . . . . . . . . . . . . . . 32
3.5 Conclusion . . . . . . . . . . . . . . . . . . . . . . . . . . . . . . . . . . . . . . . . . 35
References . . . . . . . . . . . . . . . . . . . . . . . . . . . . . . . . . . . . . . . . . . . . . 35

**4 Electrorheological Fluid-Assisted Micro-USM** . . . . . . . . . . . . . . . . 39
4.1 Introduction . . . . . . . . . . . . . . . . . . . . . . . . . . . . . . . . . . . . . . . 39
4.2 Overview of Micro-USM Process . . . . . . . . . . . . . . . . . . . . . . . 40
4.3 Chippings and Their Generation . . . . . . . . . . . . . . . . . . . . . . . . 42
4.4 Electrorheological Fluid-Assisted Micro-USM . . . . . . . . . . . . . . 42
4.5 Other Fluid-Assisted Micromachining Techniques . . . . . . . . . . . . 44
   4.5.1 Chemical-Assisted Micromachining . . . . . . . . . . . . . . . . . 44
   4.5.2 Gas-Assisted Variant of Hybrid Micromachining . . . . . . . . 45
   4.5.3 Water-Assisted Micromachining . . . . . . . . . . . . . . . . . . . . 46
4.6 Conclusion . . . . . . . . . . . . . . . . . . . . . . . . . . . . . . . . . . . . . . . . . 46
References . . . . . . . . . . . . . . . . . . . . . . . . . . . . . . . . . . . . . . . . . . . . . 47

**5 Other Assisted Hybrid Micromachining Processes** . . . . . . . . . . . . . 49
5.1 Introduction . . . . . . . . . . . . . . . . . . . . . . . . . . . . . . . . . . . . . . . 49
5.2 Vibration-Assisted Micromachining . . . . . . . . . . . . . . . . . . . . . . 50
   5.2.1 Tool Vibration Assisted Variant of Hybrid
      Micromachining . . . . . . . . . . . . . . . . . . . . . . . . . . . . . . . 50
   5.2.2 Workpiece Vibration Assisted Micromachining . . . . . . . . . 52
   5.2.3 Work Fluid Vibration Assisted Micromachining . . . . . . . . 53
   5.2.4 Objective Lens Vibration Assisted Micromachining . . . . . . 53
5.3 External Electric Field Assisted Variant of Hybrid
   Micromachining . . . . . . . . . . . . . . . . . . . . . . . . . . . . . . . . . . . . 54
5.4 Carbon Nanofibre-Assisted Micromachining . . . . . . . . . . . . . . . . 55
5.5 Conclusion . . . . . . . . . . . . . . . . . . . . . . . . . . . . . . . . . . . . . . . . . 56
References . . . . . . . . . . . . . . . . . . . . . . . . . . . . . . . . . . . . . . . . . . . . . 56

**6 Combined Variant of Hybrid Micromachining Processes** . . . . . . . . 61
6.1 Introduction . . . . . . . . . . . . . . . . . . . . . . . . . . . . . . . . . . . . . . . 61
6.2 Laser Micro-drilling and Jet Electrochemical Machining . . . . . . . 62
6.3 Micro-electrochemical Machining Combined with
   Micro-mechanical Grinding . . . . . . . . . . . . . . . . . . . . . . . . . . . . 62
6.4 Micro-electrochemical Discharge Machining . . . . . . . . . . . . . . . . 64
6.5 Simultaneous Micro-electrical Discharge Machining
   and Micro-electrochemical Machining . . . . . . . . . . . . . . . . . . . . . 65

6.6  Micro-electrical Discharge Machining Combined
     with Electrorheological Fluid-Assisted Polishing. . . . . . . . . . . . .    65
6.7  Applications . . . . . . . . . . . . . . . . . . . . . . . . . . . . . . . . . . . . . . . .    66
6.8  Conclusion . . . . . . . . . . . . . . . . . . . . . . . . . . . . . . . . . . . . . . . . .    67
References . . . . . . . . . . . . . . . . . . . . . . . . . . . . . . . . . . . . . . . . . . . .    68

# Chapter 1
# Overview of Hybrid Micro-manufacturing Processes

## 1.1 Introduction

There has been a burgeoning demand for micro-components/products in various industries with varied domains of interest such as optical, aviation, electronics, aviation, and biomedical (Nguyen 2013). The major features characterizing a micro-component or micro-product are: size ranging from few micrometers up to 100 μm, tolerances better than 1 μm, complicated 3D structure, excellent surface finish with surface roughness lesser than half μm and fabricated using multiple materials such as titanium alloys, ceramics, hard steels, etc. (Chang 2012). A number of fabrication techniques have been proposed to meet the aforementioned requirements and mass fabricate the micro-components or micro-products. The proposed techniques are based on transfer, printing, assembly, etching, and lithography (Rai-Choudhury 1997; Madou 2009). The various fabrication techniques have the potentiality to fabricate micro-sized components using inorganic as well as organic materials and lending them the desired complex 3D shapes. The micromachining techniques offer a promising approach for bridging the gap between macro and micro/nanodomains (Cardoso and Davim 2012; Piljek et al. 2014; Leondes 2007).

The use of single machining process to manufacture the micro-components has a number of limitations such as potentiality to produce the desired complex shapes with the desired accuracy, capability to mass produce such components, and predictability (Luo et al. 2005). To minimize the limitations associated with the stand-alone systems, machines with multifunctional capabilities have been developed that have the potentiality to implement several machining mechanisms on one machine. The multifunctional machining mechanism aids in rapid and economic fabrication of micro-components. Integration of conventional and nonconventional micromachining process has been the focal point of the research community recently. The integration is often referred to as hybrid micromachining processes (El-Hofy 2005). The recent applications of hybrid micromachining processes have demonstrated their capability to effectively machine hard-to-machine materials with higher geometrical

© The Author(s), under exclusive license to Springer Nature Switzerland AG 2019  
S. Bhowmik and D. Zindani, *Hybrid Micro-Machining Processes*,  
SpringerBriefs in Applied Sciences and Technology,  
https://doi.org/10.1007/978-3-030-13039-8_1

accuracy and surface integrity. The tool life has also been revealed to increase in tandem with the efficiency of the machining processes.

However, there has been no exact definition to properly describe a hybrid machining process (Zhu et al. 2013). Past researchers have given their definitions from time to time (Rajurkar et al. 1999; Aspinwall et al. 2001; Curtis et al. 2009; Lauwers 2011). As for instance, Aspinwall et al. (2001) have described hybrid machining processes as a combination of machining processes that are applied individually and independently on a single machine. They have also described the hybrid processes as assisted processes wherein two or more processes are applied simultaneously. In another definition proposed by She and Hung (2008), the machines carrying out different operations at one place are regarded as hybrid machines and the processes carried out by them are referred to as hybrid machining processes. A narrower definition was provided by Curtis et al. (2009) who have described the term "hybrid" as a machining method where two or more material removal processes work simultaneously. Following definition has been put forth by College International Pour la Recherche en productique (CIRP): the interaction of various machining mechanisms/tools/energy sources simultaneously and in a controlled manner results in hybrid machining processes. The simultaneous and controlled interaction signifies that the different mechanisms of energy/processes must act in a particular processing zone and at the same time. The hybridization therefore can have significant effect on the machining performance and the related variables.

## 1.2   Classification of Hybrid Micro-manufacturing Processes

The hybrid micromachining processes can be classified on the basis of the following:

- *Type I*: Combination of two or more processes wherein the material removal takes place because of the simultaneous action of the processes in a controlled manner. The resulting material removal is the net obtained from the individual processes.
- *Type II*: Additional usage of energy sources aiding in the net material removal, resulting in an enhanced machining performance.
- *Type III*: Usage of tools that can simultaneously machine two or more surfaces.

Electro-discharge micro-grinding (EDG) is one of the examples for the Type I category. In EDG process, material removal takes place as a combined effect of electro-discharge phase and due to abrasion resulting from grinding. The combination results in improved flushing and hence enhanced material removal rate in comparison to the electro-discharge machining. Laser-assisted micro-milling machining process falls under Type II category wherein the material is heated using laser energy which aids in enhancing the material removal rate of the milling process. The energy-assisted hybrid processes can be used for hard-to-machine materials. The processes falling under Type III category ensure that the single pass machining is sufficient for machining two or more surfaces simultaneously. A brief discussion on the classification scheme ensues next.

### 1.2.1 Compound Processes

The combination of different electrical/mechanical and chemical processes can micromachine micro-components or micro-products with the distinctive and desired benefits. The hybrid processes obtained as a result of combination can offer novel advantages over the stand-alone manufacturing systems such as elimination of tool and workpiece interaction (Rajurkar et al. 2006). Some of the recent developments on the compound processes are presented in this section.

The EDM process can be hybridized with other prominent processes to yield novel hybrid micromachining techniques such as abrasive electro-discharge micro-grinding (AEDG), electrochemical discharge micromachining, electro-discharge micro-grinding (EDG), and electrochemical discharge micro-grinding. The hybridization can overcome some of the major limitations of the stand-alone EDM process such as the significant effect on the surface integrity, increased surface roughness associated with the increased voltage and current, micro-cracks owing to thermal action, residual stresses, and microhardness change of surface layer and the subsurface.

The material removal rate in case of EDG process takes place with the rapid spark discharge between workpiece and the rotating tool. The tool and the workpiece are separated by dielectric flowing at a suitable velocity. The rotating electrode tool is what differentiates the stand-alone EDM process from the hybridized EDG process. The rotation effect provides an enhanced flushing efficiency. The improved flushing efficiency results in the ejection of molten metal from the gap between the electrode and the workpiece thereby eliminating the formation of debris. Thus, one of the major limitations of the stand-alone EDM system, i.e., the accumulation of debris within the gap is overcome with the hybridized EDG micromachining process. Researchers have carried out investigation to identify the variables affecting the EDG machining performance. Identification and selection of the suitable level for the identified factor can result in enhancing the performance parameters such as surface quality and material removal rate. Shih and Shu (2008) have recommended the use of longer pulse duration and higher peak current for achieving higher material removal rate. EDG has been used for machining of thin and fragile materials owing to the absence of mechanical forces on the workpiece from the tool giving it an upper edge over the stand-alone EDM process (Yadav and Yadava 2012).

AEDG, on the other hand, synergizes the combining effect of grinding and EDM processes wherein the metallic bonded grinding wheel replaces the graphite or metallic electrodes used in the EDG process. The synergetic combination results in increased productivity of AEDG machining process (Kozak 2002) because the material removal process takes place due to both the electro-discharge erosion and mechanical abrasion action. Some of the other connotations for AEDG process are electro-discharge abrasive grinding and electro-discharge abrasive grinding. AEDG process is being applied for machining of superhard materials, sintered carbides, metal composites, and sintered carbides (Kozak 2002). The surface finish obtained using AEDG has been revealed to be far more superior to the conventional grinding and electro-erosion grinding processes using boron nitride grinding wheels (Dąbrowski and Marciniak 2005).

Electrochemical discharge micromachining (ECDM) is another hybrid arrangement wherein the material removal takes place as a result of thermal erosion due to electrical discharges between the electrodes. The hybrid arrangement has been employed for micromachining and scribing of materials that are hard, brittle, and nonconductive in nature.

The material removal rate in ECDM process can be 10–50 times higher than ECM and EDM processes. It has been revealed that ECDM process is mainly suitable for materials that have higher tensile strength of the order of 1500 N/mm$^2$ and for alloys that are resistant to heat (Zheng et al. 2007). One of the ways in which the surface roughness of the parts obtained by ECDM process could be reduced was demonstrated by Cao et al. (2013) through the employability of micro-grinding process. The surface roughness was reduced to around 0.05 μm from few tens of micrometer with the grinding process carried out using polycrystalline diamond tools. Reduction in machining time was also observed for the hybrid process in comparison to the stand-alone conventional grinding. The versatility of the process was further investigated by Liu et al. (2013) through the investigation of surface roughness while machining of metal matrix composites. It was revealed that for grinding-assisted ECDM process, the value of surface roughness was ten times lesser than that obtained while machining the same specimen using ECDM alone.

## 1.2.2 Energy-Assisted Micromachining Processes

### 1.2.2.1 Vibration-Assisted Micromachining

A number of processes such as grinding, turning, and drilling have been assisted with vibrating tool to provide for vibration-assisted micromachining processes. The tool vibrates with a small amplitude and high frequency thereby increasing the machining performance of the stand-alone machining processes. The tool is either driven in one dimension, i.e., reciprocating or in two-dimensional space, i.e., in elliptical motion. The centroid of the vibrating tool moves in the same direction as that of the direction of cutting velocity. Owing to the appropriate combination of tool frequency, tool amplitude, and velocity of cutting, there may be instances where tool loses contact with the chip. This has positive effect as forces for machining reduces and at the same time thin chips are produced. As a result, workpiece obtained has better surface finish and dimensional accuracy relative to that obtained by employing conventional machining process (Brehl and Dow 2008). As for instance, in micro-EDM machining process wherein cylindrical tool electrode is employed, the debris flows with stability thereby providing machining stability when machining cylindrical workpiece. However, the process ends up with debris adhering to the workpiece and the tool electrode when machining rectangular shafts or holes owing to the unstable flow ability of the debris (Endo et al. 2008).

Researchers have been actively carrying our research on vibration-assisted micro-machining processes in hunt for achieving better surface finish and dimensional tolerances. Ultrasonically assisted micro-EDM process is one among such processes. Investigations have been carried out for effects of tool vibration or the vibrating workpiece on the material removal rate, quality of surface, efficiency of machining, and so on (Gao and Liu 2003). The employment of vibrating tool was revealed to enhance the circulation of dielectric fluid and hence leading to effective removal of metal from craters and improved flushing of debris. The enhanced activity significantly improved the discharge characteristics. Further, the workpiece microstructure modification and micro-cracks were minimized owing to the better discharge characteristics (Zhang et al. 2006; Shabgard et al. 2009) even at higher material rates. Significant performance improvement of micro-EDM process was reported by Cao et al. (2013) owing to the ultrasonic action and efficiency reported was up to seven times higher than the stand-alone micro-EDM process.

High aspect ratio micro-holes were revealed to be produced on tungsten carbide with ultrasonic-assisted micro-EDM process in comparison to stand-alone micro-EDM process (Jahan et al. 2010). It was reported that the material removal rate and workpiece surface quality was increased significantly. Further, a decrease in electrode wear rate with better dimensional accuracy was reported for the deep drilled micro-holes. The holes produced with ultrasonic-assisted micro-EDM process were smooth and free from defects.

Studies were also reported for ultrasonic-assisted micro-grinding processes carried out with materials such as steels (Unune and Mali 2015; Li et al. 2012), ceramics (Akbari et al. 2008), titanium (Qin et al. 2009) and nickel-based super alloys (Bhaduri et al. 2012). Better surface finish with improved tool life was reported at higher frequency of vibration and lower feed rates.

### 1.2.2.2 Thermally Assisted Micromachining

A number of materials with enhanced physical and mechanical properties have been developed in order to meet the applications in a wide range of engineering applications such as automotive, aviation, medicine, etc. Machining of such materials with conventional or stand-alone nonconventional machining processes requires higher cutting forces and therefore higher cutting temperature is generated which ultimately leads to surface and subsurface modifications. Further, the parts are required to be fabricated with higher precision and accuracy. Micromachining processes have facilitated the production of parts with high dimensional accuracy. However, the cutting forces are relatively higher at microscale and therefore the stand-alone micromachining systems also find it difficult to machine the newer materials. Higher cutting forces may result in catastrophic failure of the tool and poor surface finish.

Thermally assisted micromachining processes have aided in the machining of difficult-to-machine materials and therefore the production of parts with high aspect ratio. External source of heat results in the increasing temperature of cutting area. The increased temperature decreases the flow stress and the strain hardening rate thereby facilitating the processing of the material. The cutting forces on the tool are also reduced. Following characteristics are demanded from the external heat sources: (i) high energy density that could aid in rapid preheating of the material; (ii) reasonable cost and safety in operation relative to other stand-alone machining systems and (iii) easy location of the heating area and its size control. Plasma (Leshock et al. 2001; Shin and Kim 1996), laser (Kim et al. 2011) and oxyacetylene torches (Amin et al. 2008; Lajis et al. 2009) are few external sources of heat that have been employed in thermally assisted micromachining processes.

Laser-assisted micromachining process employs a focussed beam of laser on the materials and thereby facilitating for instantaneous heat demand for machining of materials that are difficult to be processed by the conventional machining processes. Laser has been used in tandem with milling, grinding, turning, electro-chemical machining, electro-discharge machining, etc. The laser-assisted machining processes have been reported to have higher performance efficiency relative to the stand-alone machining processes. Reduction in cutting forces and tool wear has been successfully achieved with laser-assisted micro turning on different materials such as Inconel (Garcí et al. 2013), silicon nitride (Lee et al. 2009), etc. $CO_2$ lasers have been used prominently for the laser-assisted micromachining processes. Other lasers such as neodymium-doped yttrium aluminum garnet (Nd:YAG) have also been used but their application has been limited. The studies reported have revealed reduction in cutting forces, improvement in tool life and surface integrity.

Complicated laser-assisted milling wherein laser integrated with milling has also been achieved and is employed for machining of a wide range of materials. Low power laser is used in case of laser-assisted micromachining to preheat the material locally and the material removal takes place by the milling tool. Laser-assisted micro-milling employing micro-ball end milling tool was investigated for hardened A2 tool steel (Melkote et al. 2009). The study revealed better dimensional accuracy of the microgroove produced and a lower wear rate for the tool. The surface roughness was reported to however increase with the increased cutting speed while the accuracy in depth of groove was found to be enhanced. Decreased peak resultant force has also been reported in comparison to the machining process without laser (Kumar and Melkote 2012). Development of safer and relatively more reliable, economical laser-assisted machining equipment has now become a major research area.

### 1.2.2.3  Pulse-Assisted Micromachining

The prominent example of pulse-assisted machining process is that of pulse-assisted electrochemical micromachining process (PECM) (Kock et al. 2003). PECM process has been employed effectively and economically for machining of heat resistant and high strength materials. Such materials have been machined into complex shapes such as micro-cavities, blades for micro-pumps, etc. Comparative analysis of PECM with ECM by Kozak et al. (2005) has revealed an improved machining accuracy and surface finish.

#### 1.2.2.4 Media-Assisted Micromachining

The cutting zone or the machining zone in media-assisted machining receives highly pressurized jet of water or emulsion. Highly pressurized media has ameliorated the stand-alone machining processes to enable them to cut difficult-to-cut materials such as titanium and nickel alloys. Reduction in cutting forces and effective breaking of chip has been reported with employability of highly pressurized coolant media. Furthermore, improvement in lubrication and reduction in thermal loads on cutting tools have also been accomplished with media-assisted machining processes (Courbon et al. 2009).

Turning process assisted with high pressurized coolant was used for roughing and finishing of Inconel 718, Ti6Al4V and steel. The media-assisted turning process resulted in reduction in cutting force, chip size, and tool wear owing to the hydraulic pressure between the chip and the rake face of the tool. Courbon et al. (2009) provide for more insights into the effect of cutting process onto the chip breakability, surface finish, contact length, tool temperature, and cutting forces.

Machinability of Inconel 718 was investigated by Çolak (2012) under conventional and media-assisted high pressurized cooling conditions. Cutting forces and tool flank wear was reported to decrease with the utilization of highly pressurized coolant to the cutting zone. Latest advancement has been to employ cryogenic coolant for machining and is known as cryogenic machining. The cryogenic-assisted machining process has resulted in enhanced chemical stability and improved productivity level for difficult-to-cut materials (Kenda et al. 2011; Jerold and Kumar 2013; Barletta and Tagliaferri 2006; Barletta et al. 2007).

#### 1.2.2.5 Electromagnetic-Assisted Micromachining

Application of magnetic force to the stand-alone manufacturing and machining processes has been recently exploited by the research community. The investigations carried out have reported the beneficial aspects of magnetic forces to the material removal process. Employability of magnetic abrasive particles for finishing process has been analyzed by many researchers for their reliability and feasibility. Refinement of surface for silver steel has been carried out by Khairy (2001) using finishing operation assisted with magnetic abrasive particles. Lin and Lee (2008) have reported improved machined surface with the employability of magnetic abrasive assisted micro-EDM process. Magnetic field was used with the abrasive flow machining process by Singh and Shan (2002) and improved performance was reported by their investigation.

Magnetic-assisted EDM process was first suggested for gab cleaning by De Brujin (1978). Improved circulation of debris was reported with the application of magnetic force in a direction perpendicular to the rotational force of the electrode tool. The resultant force leads to effective removal of debris from the gap which in turn results in enhanced material removal from the gap. Various investigations have suggested relatively higher aspect ratio holes produced with magnetic-assisted micro-EDM process than that produced using conventional micro-EDM machining process, with operating conditions kept the same.

### 1.2.3  Combined Hybrid Micromachining Processes

In combined hybrid micromachining processes, material removal rate is effected simultaneously by the constituting micromachining processes. The machining zone is also effected by the congregating micromachining processes. Micromachining processes with combined effect have giant potential to produce more complex shapes with enhanced material removal rate. Parts with high dimensional and surface integrity have been obtained within short production time using combined hybrid micromachining processes. Research in the field of combined hybrid micromachining processes is in nascent stage and has been done primarily on electrical and electromechanical phenomenon. Research reports have suggested lack of process capabilities to manufacture complicated 3D microstructures. Examples of such processes include simultaneous micro-EDM with micro-ECM milling (SEDCM milling), micro-ECM and micro-mechanical grinding, and micro-EDM combined with electrorheological fluid-assisted polishing and laser micro-drilling and jet electrochemical machining (JECM-LD).

## 1.3  Future Research Opportunities

Hybrid micromachining processes involve different forms of energy simultaneously that are impacted at the same machining zone. Understanding of material removal mechanism and the mechanics behind such hybrid setup is one of the future research objectives. Another future research scope is modeling and simulation of hybrid micromachining processes. Simulation of such processes can be accomplished with the combination of different methods such as finite element analysis, molecular dynamic simulation, multiscale modeling, etc. Modeling and simulation can aid in a better understanding and hence optimization of the hybrid processes.

There is requirement of specific multi-axis ultraprecision machines that possesses high stiffness (dynamic and static), high thermal stability, accurate feed drives and control mechanisms, low vibrations, low errors associated with multi-axes Interpolation, precise linear guides, high resolution static and dynamic motions, precise spindle bearings, reduced thermal effects and should be equipped with mechanism to compensate for static and dynamic positioning errors. Machines with multi-axis ultraprecision capabilities are also required for performance of tasks in diverse directions and angles. Further, smooth travel and movement of the tool call for an enhanced numerical control that will ensure high accuracies of the fabricated microstructures.

Control and continual improvement in the machining process is another important aspect that is quintessential to ensure proper implementation of the machining process in production of required microstructures. Therefore, process monitoring systems are necessary for hybrid micromachining processes that can control various variables and parameters such as chatter, cutting force, temperature, etc. Acoustic emission based systems and systems related to force and vibration signals can be employed to control and characterize the processes. However, the use of multiple sensors and other intelligent machine tools has been deemed desirable.

The precision of the hybrid micromachining processes can be enhanced further with minimized clamping and realignment tool errors. This can be ensured through the fabrication of on-machine tool and integrating the same with the hybrid micromachining processes. Another facet is the achievement of high quality of microstructures which can be ensured using noncontact, on-metrology systems.

Another future perspective of research can be the investigation into development of other novel hybrid micromachining processes that would be capable to produce free from surfaces. The goal could be achieved by considering the possibility of integration of mechanical, thermal, electrochemical, and chemical micromachining processes. Integration of single hybrid micromachining processes with other micromachining methods and assisted hybrid methods is another subject of future research that could aid in minimizing the shortcomings related to stand-alone hybrid micromachining processes.

Consideration of cost-effectiveness of a particular hybrid micromachining process over the other micromachining process is another crucial facet that needs attention from the research community. Important components aiding in calculation of cost-effectiveness such as capital investment, micromachining cost per minute, labor cost, etc., must be evaluated and investigated. Such investigations can form another future objective in the realm of hybrid micromachining processes. Investigation into cost-effectiveness can result in easy implementation and technology transfer at industrial level.

## 1.4 Conclusion

Process capabilities such as surface roughness, tool life, material removal rate, and accuracy of geometry have been improved through the employability of hybrid micromachining processes. Hybrid micromachining processes have been found to be beneficial in fabrication of complex micro-parts from hard-to-machine materials. Higher machining efficiencies have been obtained using hybrid micromachining setups. However, proper understanding of the various processes and the related parameters is quintessential to avoid any undesirable consequences such as micro-cracks and deteriorated surface finish.

Research and development of unique features in hybrid micromachining processes is just at the nascent stage. Comprehension of process mechanism, methods of simulation and modeling, development of multi-axis ultraprecise machine tools, cost-effectiveness and technology transfer to industries are among the few future works in the hybrid micromachining milieu.

# References

J. Akbari, H. Borzoie, M.H. Mamduhi, Study on ultrasonic vibration effects on grinding process of alumina ceramic ($Al_2O_3$). World Acad. Sci. Eng. Technol. **41**, 785–789 (2008)

A.N. Amin, S.B. Dolah, M.B. Mahmud, M.A. Lajis, Effects of workpiece preheating on surface roughness, chatter and tool performance during end milling of hardened steel D2. J. Mater. Process. Technol. **201**(1–3), 466–470 (2008)

D.K. Aspinwall, R.C. Dewes, J.M. Burrows, M.A. Paul, B.J. Davies, Hybrid high speed machining (HSM): system design and experimental results for grinding/HSM and EDM/HSM. CIRP Ann. Manuf. Technol. **50**(1), 145–148 (2001)

M. Barletta, V. Tagliaferri, Development of an abrasive jet machining system assisted by two fluidized beds for internal polishing of circular tubes. Int. J. Mach. Tools Manuf. **46**(3–4), 271–283 (2006)

M. Barletta, D. Ceccarelli, S. Guarino, V. Tagliaferri, Fluidized bed assisted abrasive jet machining (FB-AJM): precision internal finishing of Inconel 718 components. J. Manuf. Sci. Eng. **129**(6), 1045–1059 (2007)

D. Bhaduri, S.L. Soo, D.K. Aspinwall, D. Novovic, P. Harden, S. Bohr, D. Martin, A study on ultrasonic assisted creep feed grinding of nickel based superalloys. Proc. CIRP **1**, 359–364 (2012)

D.E. Brehl, T.A. Dow, Review of vibration-assisted machining. Precis. Eng. **32**(3), 153–172 (2008)

X.D. Cao, B.H. Kim, C.N. Chu, Hybrid micromachining of glass using ECDM and micro grinding. Int. J. Precis. Eng. Manuf. **14**(1), 5–10 (2013)

P. Cardoso, J.P. Davim, A brief review on micromachining of materials. Rev. Adv. Mater. Sci **30**(1), 98–102 (2012)

W. Chang, *Development of Hybrid Micro Machining Approaches and Test-bed* (Doctoral dissertation, Heriot-Watt University, 2012)

O. Çolak, Investigation on machining performance of inconel 718 in high pressure cooling conditions. Strojniški vestnik-J. Mech. Eng. **58**(11), 683–690 (2012)

C. Courbon, D. Kramar, P. Krajnik, F. Pusavec, J. Rech, J. Kopac, Investigation of machining performance in high-pressure jet assisted turning of Inconel 718: an experimental study. Int. J. Mach. Tools Manuf. **49**(14), 1114–1125 (2009)

D.T. Curtis, S.L. Soo, D.K. Aspinwall, C. Sage, Electrochemical superabrasive machining of a nickel-based aeroengine alloy using mounted grinding points. CIRP Ann. **58**(1), 173–176 (2009)

L. Dąbrowski, M. Marciniak, M., Investigation into hybrid abrasive and electrodischarge machining. Arch. Civ. Mech. Eng. (Oficyna Wydawnicza Politechniki Wroclawskiej), **5**(2) (2005)

H.E. De Bruijn, Effect of a magnetic field on the gap cleaning in EDM. Ann. CIRP **27**(1), 93–95 (1978)

H. El-Hofy, *Advanced Machining Processes: Nontraditional and Hybrid Machining Processes*, vol. 120 (McGraw-Hill, New York, 2005)

T. Endo, T. Tsujimoto, K. Mitsui, Study of vibration-assisted micro-EDM—the effect of vibration on machining time and stability of discharge. Precis. Eng. **32**(4), 269–277 (2008)

C. Gao, Z. Liu, A study of ultrasonically aided micro-electrical-discharge machining by the application of workpiece vibration. J. Mater. Process. Technol. **139**(1–3), 226–228 (2003)

V. Garcí, I. Arriola, O. Gonzalo, J. Leunda, Mechanisms involved in the improvement of Inconel 718 machinability by laser assisted machining (LAM). Int. J. Mach. Tools Manuf. **74**, 19–28 (2013)

M.P. Jahan, M. Rahman, Y.S. Wong, L. Fuhua, On-machine fabrication of high-aspect-ratio micro-electrodes and application in vibration-assisted micro-electrodischarge drilling of tungsten carbide. Proc. Inst. Mech. Eng., Part B: J. Eng. Manuf. **224**(5), 795–814 (2010)

B.D. Jerold, M.P. Kumar, The influence of cryogenic coolants in machining of Ti–6Al–4V. J. Manuf. Sci. Eng. **135**(3), 031005 (2013)

J. Kenda, F. Pusavec, J. Kopac, Analysis of residual stresses in sustainable cryogenic machining of nickel based alloy—Inconel 718. J. Manuf. Sci. Eng. **133**(4), 041009 (2011)

A.B. Khairy, Aspects of surface and edge finish by magnetoabrasive particles. J. Mater. Process. Technol. **116**(1), 77–83 (2001)

K.S. Kim, J.H. Kim, J.Y. Choi, C.M. Lee, A review on research and development of laser assisted turning. Int. J. Precis. Eng. Manuf. **12**(4), 753–759 (2011)

M. Kock, V. Kirchner, R. Schuster, Electrochemical micromachining with ultrashort voltage pulses—a versatile method with lithographical precision. Electrochim. Acta **48**(20–22), 3213–3219 (2003)

J. Kozak, Abrasive electrodischarge grinding (AEDG) of advanced materials. Arch. Civ. Mech. Eng. **2**(1), 83–101 (2002)

J. Kozak, D. Gulbinowicz, Z. Gulbinowicz, Investigations of MICRO electrochemical machining with ultrashort pulses, in *Proceedings of the 5th International Conference of the European Society for Precision Engineering and Nanotechnology*, Montpellier (2005), pp. 8–11

M. Kumar, S.N. Melkote, Process capability study of laser assisted micro milling of a hard-to-machine material. J. Manuf. Process. **14**(1), 41–51 (2012)

M.A. Lajis, A.K.M. Amin, A.N. Karim, C. Daud, M. Radzi, T.L. Ginta, Hot machining of hardened steels with coated carbide inserts. Am. J. Eng. Appl. Sci. **2**(2), 421–427 (2009)

B. Lauwers, Surface integrity in hybrid machining processes. Proc. Eng. **19**, 241–251 (2011)

J. Lee, S. Lim, D. Shin, H. Sohn, J. Kim, J. Kim, Laser assisted machining process of HIPed silicon nitride. JLMN-J. Laser Micro/Nanoeng. **4**, 207–211 (2009)

C.T. Leondes (ed.), *Mems/Nems: (1) Handbook Techniques and Applications Design Methods, (2) Fabrication Techniques, (3) Manufacturing Methods, (4) Sensors and Actuators, (5) Medical applications and MOEMS* (Springer Science & Business Media, 2007)

C.E. Leshock, J.N. Kim, Y.C. Shin, Plasma enhanced machining of Inconel 718: modeling of workpiece temperature with plasma heating and experimental results. Int. J. Mach. Tools Manuf. **41**(6), 877–897 (2001)

K.M. Li, Y.M. Hu, Z.Y. Yang, M.Y. Chen, Experimental study on vibration-assisted grinding. J. Manuf. Sci. Eng. **134**(4), 041009 (2012)

Y.C. Lin, H.S. Lee, Machining characteristics of magnetic force-assisted EDM. Int. J. Mach. Tools Manuf. **48**(11), 1179–1186 (2008)

J.W. Liu, T.M. Yue, Z.N. Guo, Grinding-aided electrochemical discharge machining of particulate reinforced metal matrix composites. Int. J. Adv. Manuf. Technol. **68**(9–12), 2349–2357 (2013)

X. Luo, K. Cheng, D. Webb, F. Wardle, Design of ultraprecision machine tools with applications to manufacture of miniature and micro components. J. Mater. Process. Technol. **167**(2–3), 515–528 (2005)

M. Madou, *Fundamentals of Microfabrication and Nanotechnology*, 3rd edn. (2009)

S. Melkote, M. Kumar, F. Hashimoto, G. Lahoti, Laser assisted micro-milling of hard-to-machine materials. CIRP Ann. **58**(1), 45–48 (2009)

M.D. Nguyen, *Simultaneous Micro-EDM and Micro-ECM in Low Resistivity Deionized Water* (Ph.D. thesis, National University of Singapore, 2013)

P. Piljek, Z. Keran, M. Math, Micromachining-review of literature from 1980 to 2010. Interdisc. Desc. Complex Syst.: INDECS **12**(1), 1–27 (2014)

N. Qin, Z.J. Pei, C. Treadwell, D.M. Guo, Physics-based predictive cutting force model in ultrasonic-vibration-assisted grinding for titanium drilling. J. Manuf. Sci. Eng. **131**(4), 041011 (2009)

P. Rai-Choudhury, *Handbook of Microlithography, Micromachining, and Microfabrication*, vol. 1: Microlithography: SPIE Opt (1997)

K.P. Rajurkar, D. Zhu, J.A. McGeough, J. Kozak, A. De Silva, New developments in electro-chemical machining. CIRP Ann. **48**(2), 567–579 (1999)

K.P. Rajurkar, G. Levy, A. Malshe, M.M. Sundaram, J. McGeough, X. Hu, R. Resnick, A. DeSilva, Micro and nano machining by electro-physical and chemical processes. CIRP Ann. Manuf. Technol. **55**(2), 643–666 (2006)

M.R. Shabgard, B. Sadizadeh, H. Kakoulvand, The effect of ultrasonic vibration of workpiece in electrical discharge machining of AISIH13 tool steel. World Acad. Sci. Eng. Technol. **3**, 332–336 (2009)

C.H. She, C.W. Hung, Development of multi-axis numerical control program for mill—turn machine. Proc. Inst. Mech. Eng., Part B: J. Eng. Manuf. **222**(6), 741–745 (2008)

H.R. Shih, K.M. Shu, A study of electrical discharge grinding using a rotary disk electrode. Int. J. Adv. Manuf. Technol. **38**(1–2), 59–67 (2008)

Y.C. Shin, J.N. Kim, Plasma enhanced machining of Inconel 718, in *ASME International Mechanical Engineering Congress and Exposition*, Atlanta, vol. 4 (1996), pp. 243–249

S. Singh, H.S. Shan, Development of magneto abrasive flow machining process. Int. J. Mach. Tools Manuf. **42**(8), 953–959 (2002)

D.R. Unune, H.S. Mali, Current status and applications of hybrid micro-machining processes: a review. Proc. Inst. Mech. Eng., Part B: J. Eng. Manuf. **229**(10), 1681–1693 (2015)

R.N. Yadav, V. Yadava, Electrical discharge grinding (EDG): a review, in *Proceedings of the National Conference on Trends and Advances in Mechanical Engineering*, YMCA University of Science & Technology, Faridabad, Haryana (2012), pp. 590–597

Q.H. Zhang, R. Du, J.H. Zhang, Q.B. Zhang, An investigation of ultrasonic-assisted electrical discharge machining in gas. Int. J. Mach. Tools Manuf. **46**(12–13), 1582–1588 (2006)

Z.P. Zheng, K.L. Wu, Y.S. Hsu, F.Y. Huang, B.H. Yan, Feasibility of 3D surface machining on pyrex glass by electrochemical discharge machining (ECDM), in *Proc. AEMS07* (2007), pp. 28–30

Z. Zhu, V.G. Dhokia, A. Nassehi, S.T. Newman, A review of hybrid manufacturing processes—state of the art and future perspectives. Int. J. Comput. Integr. Manuf. **26**(7), 596–615 (2013)

# Chapter 2
# Laser-Assisted Micromachining

## 2.1 Introduction

A number of unique properties are associated with laser radiations such as high monochromaticity, spatial coherence, temporal coherence, and electromagnetic energy flux. The laser has the potentiality to travel as a very narrow beam and it is highly directional due to high temporal and spatial coherence. This makes it possible to direct the laser upon the desired small area with very high radiance (Rykalin et al. 1978). Laser when used as a direct source of energy results in deposition, removal, or alteration of material properties (Sugioka et al. 2010). Efficient control of the depth of penetration depth and the amount of energy to be dispersed to the desired area are the few advantages of using laser. Characteristics such as low heat input, high lateral resolution, and high flexibility make laser suitable for microtechnology.

Long-pulsed lasers have been employed since long for the creation of desired features on a substrate. Some of the commonly used laser beam machining processes are laser drilling, laser grooving, laser turning, and laser cutting. Mechanism of plasma formation and optical absorption are the supporting base of laser which often results in localized deposition of heat, formation of micro-cracks and hence damage to the area surrounding the machining area. Laser beam micromachining, on the other hand, utilizes the properties of ultrashort lasers to generate micro-features on the substrates by acquiring an exceptional degree of control. The properties acquired by the laser beam aids in minimizing damage to the surroundings. Multiphoton nonlinear optical absorption is responsible for deposition of laser energy into small volumes. The optical absorption phenomenon is followed by avalanche ionization. Time scale for heat transport ranges several nanoseconds to few microseconds whereas time frame range for electron–phonon coupling is picosecond to nanosecond. The damage to the surrounding is avoidable if deposition of laser energy takes place for a time scale that is much shorter than electron–phonon coupling and heat diffusion rate. A number

© The Author(s), under exclusive license to Springer Nature Switzerland AG 2019
S. Bhowmik and D. Zindani, *Hybrid Micro-Machining Processes*,
SpringerBriefs in Applied Sciences and Technology,
https://doi.org/10.1007/978-3-030-13039-8_2

of factors that decides on the feature size are beam quality, f-number of lens and wavelength (Meijer 2004). LBMM is now being employed by industries in variety of engineering fields such as micro-chemistry, micro-biology, micro-electronics, and micro-optics. Some of the micro-sized features that are being fabricated using LBMM are read-only memory chips, miniaturized photonic components, optical data memory, hollow channel waveguides in communication network.

Capabilities of LBMM process has been stretched to fabricate products with much more precision and higher dimensional tolerances. LBMM process is now integrated with other standalone micromachining processes that have resulted in economical production of parts with high quality. This chapter briefs on some of the laser-assisted hybrid micromachining processes. However, before proceeding for the same, basic concepts regarding LBMM process has been discussed in the next section.

## 2.2 Conceptual Framework to Laser Beam Micromachining (LBMM)

Gas- or solid-based lasers are the main sources for both the short and ultrashort lasers. A laser can be referred to as ultrashort depending on whether the thermal diffusion depth is smaller than the optical penetration depth. If thermal diffusion rate is smaller than the optical penetration depth, laser is known as ultrashort laser (Meijer 2004). LBMM process, depicted in Fig. 2.1, employs a wide range of lasers providing wavelengths from deep ultraviolet to mid-infrared. The examples of lasers in the UV range are Nd:YAG, Nd:YLF, and Nd:YVO$_4$.

For micro-fabrication, laser ablation has been considered as one of the most efficient mechanical machining methods. Intense laser radiation results in ablation of

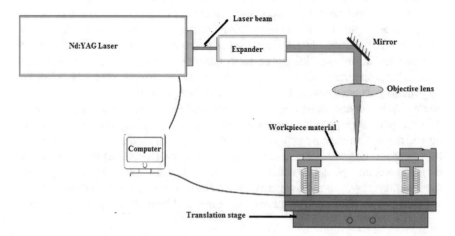

**Fig. 2.1** Process outline of laser beam micromachining

the workpiece surface. Nanostructures are obtained when the constituents of the target material get ejected due to the intense laser radiations. Ablation rate or the rate of material removal is typically around one-tenth of the monolayer per laser pulse. Linear absorption forms the major part of the main absorption phenomenon when the pulse width associated with the laser is long whereas at short laser pulse width it is the nonlinear absorption phenomenon that forms the major part of the main absorption.

The intensity of laser with the depth of the target material depends on the coefficient of absorption ($\alpha$) of the material which in turn depends on the wavelength and temperature of the material. However, it is Beer–Lambert law that dictates the decay of laser intensity for constant $\alpha$. The intensity of the laser according to Beer–Lambert law is given by $I(z) = I_0 e^{-\alpha z}$. When the value for laser intensity drops to $1/e$ of the initial value of intensity, then it is called as absorption depth ($\delta$), i.e., $\delta = 1/\alpha$. The released radiations from the laser are either absorbed by the vibrational transitions in atoms or by exciting free electrons. Electronic excitations taking place in ions, atoms, or molecules is another medium through which the laser radiations are absorbed.

In case of metals, intra- and inter-band transitions are responsible for electronic excitations. Dielectric constant $\xi$ encompasses the summation obtained from contribution due to lattice vibrations ($\xi^L$), interband transitions ($\xi^D$), and intraband transitions. Laser excitations entail the participation of both conduction-band electrons as well as valence-band electrons. Free electron forms the medium of participation for the conduction-band electrons whereas interband response makes possible the participation of valence-band electrons. The interaction between photons and electrons takes place within the conduction band. Intraband absorption results in further increase in energy of the conduction-band electrons and the phenomenon of Inverse Bremsstrahlung is responsible for this.

Free electrons dominate the phenomenon of optical absorption in case of metals. Transfer of energy takes place through subsequent collisions of phonons (Yao et al. 2005). The dominance of free electrons over the optical absorption exists till the frequency is well below the plasma frequency ($\omega_p$) that relates the optical properties of the metal to the electron density. As the plasma frequency increases, electrons find it difficult to screen through the electric field of the light and therefore the contribution of free electrons decreases. Small depth of absorption may be attributed to the large absorption coefficient in metals. The depth of absorption is referred to as skin depth.

Several mechanisms are involved in case of laser ablation of polymeric materials. Mechanisms such involved may be photochemical, photophysical, photothermal, or combination of any of these mechanisms. Bond breaking in case of photochemical mechanism takes place as a result of electronic excitation. In case of photothermal ablation, thermal bond breaking takes place as a result of thermalized electronic ablation. Pivotal role is played by both the thermal and nonthermal processes in case of photophysical mechanism.

The dielectric function in case of insulators and semiconductors has contribution from electrons as well as ions. Within the nonmetallic solids, refractive component of dielectric function dominates when the frequency is far from electronic resonance whereas near the resonance it is the absorptive component that dominates. Electronic conduction does not take place in case of insulators because of nonavailability of free carriers in the band of conduction. Therefore electron–hole pairs are created on the absorption of optics in insulators and semiconductors. Absorption of photons with energy lesser than the bulk bandgap energy of the material will not occur. Bulk bandgap energy is the energy that is required by the electrons existing in the highest energy state to cross the gap and reach the lowest level of energy in the band of conduction. Similar optical properties have been revealed to be possessed by semiconductors and insulators that are above and below their bandgap energies.

Single-photon interband transitions dominates the solid–radiation interaction phenomenon when the photon energy $h\upsilon$ is greater than the bulk bandgap energy. Electronic transitions and excitations arising due to vibrations within the lattice dictates the optical properties of the material in case the photon energy is less than the bandgap energy of the material. However, it has been revealed that interaction between laser radiation and insulators is enhanced with the presence of impurities and defects.

In case of dielectric materials, strong field ionization generates free electrons. Collision between electrons takes place once the free electrons is generated. As a result of collisions avalanche or impact ionization occurs. However, large band gap is possessed by dielectric materials and therefore to deposit the requisite energy, optical breakdown process is required. Further, the utiliation of nonlinear absorption is quintessential owing to the low linear light absorption. Deposition of the laser energy takes place in the focal volume and then the ionization of the material takes place because of the high intensity associated with the ultrashort laser. The transformation of material in the focal volume to absorbing plasma occurs due to the laser-induced breakdown. Factors such as ionization potential, laser pulse wavelength, and pulse duration affect the breakdown threshold (Fan and Longtin 2001). The transformation process converts optically transparent material to opaque material thereby restricting the incoming laser. As a result of positive charge created by ultrashort laser in and around the vicinity of the dielectric material, Coulomb explosion with very high kinetic energy takes place. Smooth surface at atomic levels is generated because of the traveling of fast ions generated as a result of Coulomb explosion. The mechanism forms the basis for processing of dielectric material given the width of laser pulse being in sub-picosecond range (Ashkenasi et al. 2003).

The above discussion and the theory are also applicable for ceramics when the laser intensity is very high. This is because at lower intensity, scattering of laser irradiation takes place.

Electrical breakdown is the main cause of material removal in case of materials that are optically transparent. The relative strength of the electric field at the exit surface in relation to that existing at entrance surface is greater than unity. A phase shift of 180° occurs at the entrance whenever transmission of laser takes place from lesser optically dense medium to higher dense medium. However, phase shift is absent at the exit. The ratio $(E_e/E_t) = \{(2n)/(n + 1)\}$ calculates the relative electric field strength at exit and at the entrance and "$n$" is the refractive index of the medium. Further, the relative intensity of the laser at the exit surface and the entrance surface is given by $(I_e/I_t) = (E_e/E_t)^2$. Therefore it is easier to process a material from the rear surface for any material with refractive index greater than one.

## 2.3 Mechanism and Principle of Laser Ablation

One of the most efficient mechanical machining methods for fabrication at micro-level is that of laser ablation. Intense laser radiation is the main source for ablation of a workpiece surface. Intense energy from laser results in the ejection of constituents from the target material and thereby formation of nanostructures. Modification of the target surface or change in composition takes place at mesoscopic levels. The ablation process initiates above certain threshold fluence. The threshold influence depends on a number of factors such as the mechanism of absorption, properties of workpiece material, defects in workpiece material, surface morphology, and laser parameters such as pulse duration and its wavelength. Typical values of threshold fluence vary from material to material as for instance it lies between 1 and 10 J/cm$^2$ for metallics and for materials that are organic its between 0.1 and 1 J/cm$^2$. The ablation threshold may reduce in case of multiple laser pulses being used for laser ablation. The reason may be attributed to the incubation effect. The relationship between multiphase threshold and single pulse threshold can be established using $F_{th}(N) = F_{th}(1)N^{\xi-1}$ where degree of incubation is denoted by $\xi$. There is no incubation effect for $\xi = 1$. Incubation effect is greatly influenced by defects that facilitates for resonant multiphoton transition.

The volumetric material removal rate per pulse shows a logarithmic increment with the fluence above the ablation threshold. This could be established in accordance with the Beer–Lambert law. A number of parameters such as wavelength, pulse length, fluence, etc., are responsible for the phenomenon of laser ablation. The photothermal mechanism is mainly responsible for the ablation process for situations of low fluence. At low fluence material evaporation and sublimation forms the major material removal base. Normal boiling takes place at low fluence that results from heterogeneous nucleation of generated vapor bubbles. However, explosive boiling may occur when the material reaches its thermodynamic critical temperature. Non-thermal photochemical ablation takes place when the time of excitation is shorter than the time of thermalization process.

The plume ejected from the irradiated zone is responsible for the material removal process in LBMM process. Ablated material is transformed into plasma at high laser intensities in which case the electric field of laser pulse exceeds the optical breakdown threshold value. Extreme nonequilibrium conditions may arise for laser with ultra-short pulses. In the subsequent subsection, discussion on laser ablation mechanism with different duration pulses, i.e., nanosecond, picosecond, and femtosecond.

### 2.3.1 Laser Ablation Process Through Nanosecond Laser

The photothermal process in case of nanosecond laser ablation process results in generation of nanoparticles. Heating of the target material takes place as a combinatorial effect of electronic and vibrational excitations of the workpiece. At lower values of fluence, thermal penetration depth determines the extent to which laser energy gets absorbed by the target material. Thermal evaporation dominates the mechanism of laser ablation in the region of thermal penetration. Ionization of laser plumes occurs in case of laser fluence that is close to that of its threshold value. Direct heating of laser is responsible for the ionization process. However, optical breakdown is responsible for ionization phenomenon in situations where the intensity of laser is higher than that of gas ionization. The phenomenon of phase explosion takes place when the irradiance of laser is greater than $10^{11}$ W/cm$^2$. The phenomenon is active wherein the temperature of the target surface becomes equal to that thermodynamic critical point. In phase explosion, transformation of matter to mix of liquid and vapor droplets takes place from the state of overheated liquid. Ejection of molten droplets then takes place at supersonic velocity as a result of generation of dense plasma possessing higher pressure and temperature at the end of laser plume (Sugioka et al. 2010). The topography of the target material, however, gets modified at the surrounding ablated region owing to the re-solidification of the ejected molten liquid material into films.

### 2.3.2   Laser Ablation Process Through Picosecond Laser

Critical point phase separation phenomenon is the primary material removal source during picosecond laser ablation process. The process does not lead to density change of the target material until maximum temperature is reached. Once the maximum temperature is reached decrease in density is observed and at this stage, the ablation process initiates for the material region entering the unstable zone. The region of the target material with lower mass will condense back on the target surface. Equations of heat diffusion can be used for the determination of temperature separation serving as a threshold for initiation of ablation process. Electrons are the first to absorb the laser energy during the ablation process by laser. The absorbed energy is then transferred to lattice via electron–phonon coupling. Solid-plasma and solid–vapor transitions occur at the surface whereas inside the metal, metal is present in liquid phase.

### 2.3.3   Laser Ablation Process Through Femtosecond Laser

There is the absence of precise threshold for laser-induced damage in cases of lasers with pulses longer than 10 ps. The fact may be attributed to the uneven distribution of surface carriers. Multiphoton ionization process frees the bound electrons in case of laser pulses lesser than 1 ps. In this case, the threshold value for laser-induced damage is precisely defined (Liu et al. 1997). Rapid thermalization takes place within the electron gas when laser of femtosecond order strikes the workpiece material. As a result of this process, scattering of electron–electron occurs which then leads to a distribution of Fermi–Dirac temperature. Lattice shape is induced on the surface of target material owing to the energy possessed by the electrons created by the thermalization process. Lattice shape formed is eliminated from the target material surface because of lattice bond breaking and expansion of material that releases energy (Chung et al. 2009). Transfer of heat from electrons to the lattice takes place as a result of electron–phonon coupling. Higher the heat transfer rate the greater is the coupling between electron and phonon. In case of femtosecond lasers, mechanical relaxation time is much greater than the deposition time of energy. The mechanical relaxation time, $\tau_m$ can be calculated using $(L_p + L_e)/C_M$, where depth of optical penetration is denoted by $L_p$ and the depth of electron energy transfer is indicated by $L_e$ and speed of rarefraction wave in the target material is denoted by $C_M$. Heating of the target material surface takes place at a fast pace if the pulse duration is shorter. This is because the time taken to start the collective motion of the atom within the absorbed volume is very short. As a result of fast surface heating, high temperature is reached with density remaining unchanged. The transition of solid to vapor takes place directly as a result of absorption of energy at the top of the target material surface. The dissipated heat does not get out of the spot radius of the beam because of the shorter duration of the laser pulse (Sugioka et al. 2010).

Avalanche and photon ionization processes are responsible for absorption of femtosecond laser. Excitations of electrons take place as a result of collision between photons and electrons and forms the major core of the photon ionization process. Multiphoton ionization process occurs when the intensity of the laser is above $10^{13}$ W/cm$^2$. In this process, several photons with similar wavelengths and energy strike the bounded electrons. However, tunneling frees the electrons in case the laser intensity is more than $10^{15}$ W/cm$^2$. The process of avalanche ionization, on the other hand, involves the collision between free and laser-heated electrons generated by the multiphoton ionization with that of the electrons from valence band leading to the generation of more free electrons. Formation of plasma marking the inception of ablation process takes place when the density of free electrons reaches its critical value.

The volumetric ablated material using laser depends to a great extent on the area of focussed laser and the penetration level of the laser heat. Fine resolution of machining can be accomplished with the volume of material ablated by the laser pulse. The diameter of the laser spot is directly dependent on the laser wavelength, focal length of the lens, and on the square of beam quality whereas it is inversely proportional to the diameter of the entering laser beam. Therefore, short wavelength of laser irradiation results into fine resolution machining. Short wavelength also produces uniform processing efficiency (Liu et al. 1997). The rate of ablation for laser with low fluence is given by Eq. (2.1)

$$L = \alpha^{-1} \ln\left(\frac{\varphi}{\varphi_{th}^\alpha}\right) \tag{2.1}$$

Equation (2.2) on the other hand provides ablation rate for lasers with high fluence

$$L = l \ln\left(\frac{\varphi}{\varphi_{th}^{l}}\right) \tag{2.2}$$

where
$\alpha^{-1}$ is optical penetration depth, $\varphi_{th}^{\alpha}$ and $\varphi_{th}^{l}$ are the threshold values of fluence.

## 2.4  Laser-Assisted Variant of Hybrid Micromachining Process

Newer materials with improved strength and thermal properties are often difficult to be fabricated using standalone conventional and nontraditional processes of machining. Low thermal conductivity and higher tensile strength result in higher cutting forces and temperature which ultimately leads to shorter tool life. Therefore research trend has been focussed widely on integrating two or more machining processes in order to provide for assisted machining processes both at micro- and macrolevel. Laser-assisted micromachining is one such hybrid micromachining process on which number of studies have been performed. The majority of work has been on the parameters such as machining forces and surface roughness. Subsequent subsections provides a brief insight into different laser-assisted micromachining processes.

### 2.4.1  Laser Beam Assisted Water Jet Micromachining

The material removal process does not take place by melting and vapourisation of the workpiece material but rather by first softening the target material through laser heating and then removing the material from the softened target material by high-pressure water jet. Usage of water jet reduces the temperature at the cutting zone of the material and thereby resulting in lesser thermal damages. The energy associated with the laser required for material removal is also reduced and therefore laser could be used with higher transverse speed. However, the hybrid process associates with itself the disadvantage of thermal shock to the target material which may result in micro-cracks within the workpiece/target material. The contribution of the thermal shock on the workpiece material can, however, be reduced with the usage of suitable waterjet pressure. Simultaneous motion of the laser beam and waterjet is quintessential for the proper removal of laser softened workpiece material. The hybrid process, therefore, comes equipped with the hybrid laser–waterjet cutting head. The waterjet nozzle is placed beside laser beam within this hybrid arrangement. Laser-assisted waterjet process has reported a higher rate of material removal without the formation of HAZ in comparison to other standalone conventional and nonconventional machining processes (Tangwarodomnukun 2012; Tangwarodomnukun et al. 2012).

**Fig. 2.2** Schematic of laser beam assisted jet electrochemical micromachining

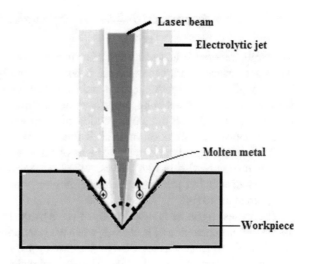

## 2.4.2 Laser Beam Assisted Jet Electrochemical Micromachining

Achievement of better localization is the primary objective of employing laser with electrochemical machining. Better localization leads to better precision. Laser energy thermally activates the outer layer of the target material. Electrochemical dissolution forms the main material removal process in this hybrid micromachining setup. The role of laser is to properly direct and assist the electrochemical energy. Employment of laser enhances the kinetics of electrochemical reactions and hence localization of dissolution to the desired area. The dimensional precision has been revealed to improve with the hybrid micromachining setup, depicted in Fig. 2.2, since the process of material removal takes place in the axial direction instead of taking place in the lateral direction.

Apart from producing effect of localization, laser beam also aids in the elimination of surface oxide layer from the workpiece material. This ultimately results in electrochemical machining of workpiece material such as titanium in the absence of hazardous electrolyte. Experiments with stainless steel, titanium, hastealloy, and aluminum alloy have proved improved accuracy and productivity of the laser beam assisted waterjet micromachining process over the standalone jet electrochemical machining. Noticeable improvements in shape and surface roughness have also been reported for holes and cavities (De Silva et al. 2004, 2011; Pajak et al. 2004, 2006).

### 2.4.3  Laser Beam Assisted Micro-milling/Grinding Process

The target material is heated by laser beam with low power and thereby softening the target surface. Workpiece material is heated intensely at the desired local area. The strength of the target material is reduced resulting in improved machinability. Reduced forces of machining, improved surface finish, and machining accuracy has been obtained with the employability of laser-assisted micro-milling/grinding process. However, formation of HAZ is unavoidable with the machining process. With the increasing temperature, the density of cracks also increases (Lauwers 2011). High temperature also has an adverse effect on the tool life. Further, accelerated diffusion–dissolution process also results in premature degradation of the cutting tool (Sun et al. 2010).

The process response is significantly affected by the laser variables. In laser beam assisted micromachining of H13 steel, when the laser power was increased from 0 to 10 W, the thrust force was revealed to decrease by 15%, whereas surface roughness showed an increment of nearly 35%. The dimensional accuracy of the groove depth was found to increase with the laser heating. No HAZ was reported after micro-cutting for the material temperature below the critical range (Singh 2007; Singh and Melkote 2007; Singh et al. 2008). For A2 tool steel, decrease in peak force, tool wear and increased accuracy in groove geometries have been reported (Kumar 2011; Melkote et al. 2009; Kumar and Melkote 2012; Kumar et al. 2012). However, one of the major disadvantages associated with the laser-assisted micro-milling/grinding is that of increased burr height. The burr height is much smaller in case of tool diameter being greater than the spot size of the laser. Increased surface roughness has been revealed with laser spot size being greater than the diameter of the tool (Kumar 2011).

## 2.5  Conclusion

This chapter elucidates on the process mechanism of ultrafast laser ablation process. The main experimental parameters involved in the laser beam micromachining process have been identified. A brief discussion has been provided on the mechanisms of optical breakdown that includes phenomenon of multiphoton absorption and avalanche ionization. Further, the quality of the micromachined components depends on polarization and plasma interaction. In order to have more focussed laser beam ultrashort lasers of different orders viz-a-viz femtosecond, picosecond, and nanosecond have also been discussed briefly.

Minimization of the laser-induced effects and optimization of energy coupling results in high degree of process control for the laser beam micromachining process. The chemical composition and the properties associated with ablated particles could be controlled with the employability of temporal pulse forming. To aid in achieving the goal of machining difficult-to-machine materials with high-dimensional accuracy and tolerances, laser has been integrated with the other micromachining processes and the results obtained have found to be encouraging.

# References

D. Ashkenasi, G. Müller, A. Rosenfeld, R. Stoian, I.V. Hertel, N.M. Bulgakova, E.E.B. Campbell, Fundamentals and advantages of ultrafast micro-structuring of transparent materials. Appl. Phys. A **77**(2), 223–228 (2003)

I.Y. Chung, J.D. Kim, K.H. Kang, Ablation drilling of invar alloy using ultrashort pulsed laser. Int. J. Precis. Eng. Manuf. **10**(2), 11–16 (2009)

A.K.M. De Silva, P.T. Pajak, D.K. Harrison, J.A. McGeough, Modelling and experimental investigation of laser assisted jet electrochemical machining. CIRP Ann. Manuf. Technol. **53**(1), 179–182 (2004)

A.K.M. De Silva, P.T. Pajak, J.A. McGeough, D.K. Harrison, Thermal effects in laser assisted jet electrochemical machining. CIRP Ann. Manuf. Technol. **60**(1), 243–246 (2011)

C.H. Fan, J.P. Longtin, Modeling optical breakdown in dielectrics during ultrafast laser processing. Appl. Opt. **40**(18), 3124–3131 (2001)

M. Kumar, *Laser Assisted Micro Milling of Hard Materials* (Doctoral dissertation, Georgia Institute of Technology, 2011)

M. Kumar, S.N. Melkote, Process capability study of laser assisted micro milling of a hard-to-machine material. J. Manuf. Process. **14**(1), 41–51 (2012)

M. Kumar, S.N. Melkote, R. M'Saoubi, Wear behavior of coated tools in laser assisted micro-milling of hardened steel. Wear **296**(1–2), 510–518 (2012)

B. Lauwers, Surface integrity in hybrid machining processes. Proc. Eng. **19**, 241–251 (2011)

X. Liu, D. Du, G. Mourou, Laser ablation and micromachining with ultrashort laser pulses. IEEE J. Quantum Electron. **33**(10), 1706–1716 (1997)

J. Meijer, Laser beam machining (LBM), state of the art and new opportunities. J. Mater. Process. Technol. **149**(1–3), 2–17 (2004)

S. Melkote, M. Kumar, F. Hashimoto, G. Lahoti, Laser assisted micro-milling of hard-to-machine materials. CIRP Ann. **58**(1), 45–48 (2009)

P.T. Pajak, A.K.M. De Silva, J.A. McGeough, D.K. Harrison, Modelling the aspects of precision and efficiency in laser-assisted jet electrochemical machining (LAJECM). J. Mater. Process. Technol. **149**(1–3), 512–518 (2004)

P.T. Pajak, A.K.M. Desilva, D.K. Harrison, J.A. Mcgeough, Precision and efficiency of laser assisted jet electrochemical machining. Precis. Eng. **30**(3), 288–298 (2006)

N.N. Rykalin, A.A. Uglov, Kokora, A., *Laser Machining and Welding* (Pergamon, 1978)

R.K. Singh, *Laser Assisted Mechanical Micromachining of Hard-to-Machine Materials* (Doctoral dissertation, Georgia Institute of Technology, 2007)

R. Singh, S.N. Melkote, Characterization of a hybrid laser-assisted mechanical micromachining (LAMM) process for a difficult-to-machine material. Int. J. Mach. Tools Manuf. **47**(7–8), 1139–1150 (2007)

R. Singh, M.J. Alberts, S.N. Melkote, Characterization and prediction of the heat-affected zone in a laser-assisted mechanical micromachining process. Int. J. Mach. Tools Manuf. **48**(9), 994–1004 (2008)

K. Sugioka, M. Meunier, A. Piqué (eds.), *Laser Precision Microfabrication*, vol. 135 (Springer, 2010)

S. Sun, M. Brandt, M.S. Dargusch, Thermally enhanced machining of hard-to-machine materials—a review. Int. J. Mach. Tools Manuf. **50**(8), 663–680 (2010)

V. Tangwarodomnukun, *Towards Damage-free Micro-fabrication of Silicon Substrates using a Hybrid Laser-Waterjet Technology* (Doctoral dissertation, Ph. D. dissertation, The University of New South Wales, 2012)

V. Tangwarodomnukun, J. Wang, C.Z. Huang, H.T. Zhu, An investigation of hybrid laser–waterjet ablation of silicon substrates. Int. J. Mach. Tools Manuf. **56**, 39–49 (2012)

Y.L. Yao, H. Chen, W. Zhang, Time scale effects in laser material removal: a review. Int. J. Adv. Manuf. Technol. **26**(5–6), 598–608 (2005)

# Chapter 3
# Magnetic Field Assisted Micro-EDM

## 3.1 Introduction

One of the most emerging and promising methods for manufacturing complex three-dimensional structure with high level of accuracy is electrical discharge machining (EDM). Material removal in EDM process is through a series of sparks produced between the electrodes. As a result, intense heat is generated at the workpiece surface leading to melting and vaporization of the metal (Jain 2010). The EDM material removal process is, therefore, a noncontact mechanism and has the potentiality to machine almost all materials that are electrically conductive regardless of their strength, toughness, and hardness. Shaping of complex structures has been achieved with surface roughness as close as $R_z = 0.4\,\mu\text{m}$ (Kunieda et al. 2005). EDM process has been used for a wide range of applications such as manufacturing of dies and molds, drilling of holes with burr-free surfaces, critical components in automobile, medical, and aerospace industries (Shao and Rajurkar 2015).

The proved versatility of the EDM process has attracted the attention of the scientific community to explore its applicability at microscale giving rise to micro-EDM. Micro-EDM process has been used to produce cutting tools and micro-molds from difficult to machine materials, and also difficult to shape structures such as fuel injection nozzles, parts for micro-mechatronic actuators, spinneret holes for processing and manufacturing of synthetic fibers and tools with micro dimensions for producing micro-sized features (Rajurkar et al. 2006). The process has even been extended to the machining of semiconducting materials such as silicon (Song et al. 2000) and ceramics with insulating characteristics such as $Si_3N_4$ (Muttamara et al. 2003). This has been achieved by introduction of secondary conductive phase such as WC, $TiB_2$, TiN, etc., inclusion of such conductive phases has also made it possible to machine ceramics based on $ZrO_2$, $Al_2O_3$ that possesses excellent physical and mechanical properties (Kunieda et al. 2005; Lauwers et al. 2004).

The material removal process in micro-EDM process comprising of melting and vaporization process results in resolidification of molten material on the electrode itself and the formation of debris. Removal of debris from the machining zone is required to eliminate generation of secondary sparking which may otherwise result in damage to the workpiece surface and reduced machining efficiency. Therefore, it is essential to place an effective debris removal mechanism. The commonly used flushing techniques in conventional EDM process such as jet and water flushing cannot be used for micro-EDM due to the constraint of size. Further, the forces from flushing may not be sustained by the fragile microelectrodes. Thus, external flushing is avoided in case of micro-EDM. The debris removal process, therefore, depends on the force provided by the rotating electrode, i.e., the centrifugal force of the electrode tool (Ghoreishi and Atkinson 2002). Number of attempts have been made to propose an effective debris removal mechanism such as ultrasonic vibration (Kremer et al. 1989; Yeo and Tan 1999; Yan et al. 2002; Wansheng et al. 2002; Gao and Liu 2003) which has been reported to decrease in machining efficiency. Modification of electrode shape is another attempt made by researchers for effective debris removal (Yan et al. 1999) which has, however, resulted in unbalanced rotation and vibration of electrodes. Some of the current research in this field has been the employability of planetary motion to the electrode and gravitational force (Yu et al. 2002; Murali and Yeo 2004).

Present chapter illuminates the readers on the usage of magnetic field as an effective mechanism for removing the debris from the machining zone. The chapter has been organized into following sections: difference between macro and micro-EDM process with due consideration to working principle, mechanism of micro-EDM and then the basics of magnetic-assisted micro-EDM process has been elaborated. The chapter finally ends with the concluding remarks.

## 3.2  Electrical Discharge Machining (EDM) and Micro-EDM Process

Electrical discharge machining (EDM) is a thermoelectric process wherein material removal tales place through sparks that melts and vaporizes the material. The spark is produced between a tool and the workpiece. As the gap between the tool and the target material reduces, the applied voltage crosses its threshold value which results in generation of spark. Once the sparking takes place, there occurs increment in the gap between the electrode tool and the workpiece. This results in setting up of new spark at the next nearest point between the electrodes. Continuous sparking is accomplished by the in-place control mechanism of inter-electrode gap maintenance. Melting and vaporization form the major part of the material removal mechanism whereas the liquid dielectric medium removes the rest. Even after the completion of the material removal process some of the material resolidifies on the surface of the target material. Craters are created on the tool as well as on the surface of the target material (Allen and Lecheheb 1996).

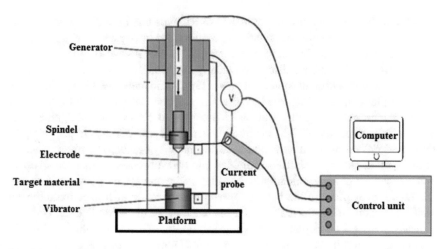

**Fig. 3.1**  Micro-EDM process setup

Spark generation and the subsequent material removal mechanism could be considered as stochastic processes. The material removal mechanism for micro-EDM is identical to that of EDM process. The process outline of micro-EDM process is delineated in Fig. 3.1. Only the differentiating part is that of precision servo system that has high sensitivity and better accuracy in dimensions and geometry, i.e., of the order of $\pm 0.4$ μm (Jahan 2013). Keeping discharge energy very low is quintessential for obtaining higher machining accuracy in micro-EDM process. Lower capacitance and open circuit voltage results in achieving minimized discharge energy in resistance—capacitance (RC) circuit. However, it is not possible to set the value of capacitance lesser than the stray capacitance of the electrical circuit in a RC circuit. Thus, achieving of lower discharge energy is possible with low open circuit voltage. Decreasing the current and/or pulse on times also results in reducing material removal per discharge of the spark (Mahendran et al. 2010) which leads to more precise control over the workpiece dimension. Small inter-electrode gaps are achievable in micro-EDM process with reduced gap voltages. Expansion of plasma channel is curtailed as a result of reduced discharge current and therefore results in much smaller diameter of the plasma channel in micro-EDM in comparison to macroscale EDM (Mahendran et al. 2010). Better surface finish and dimensional accuracy as well as generation of small diameter debris have been achieved with the use of RC circuits (Jahan et al. 2009). However, low duty cycle is one of major disadvantage of the RC circuits.

Drawback of low duty cycle in RC circuits have been done away with the traditional transistor-based circuits, i.e., transistor–resistor and transistor–inductor types. More fluctuation in voltage has been reported for transistor–resistor discharge circuit in comparison to RC-type power supply. On the other hand, an increasing tendency of spark gap voltage is shown by the transistor–inductor type circuit of discharge. In RC circuit, the capacitor discharges instantaneously after the gap breakdown and also there is a drop in the gap voltage much below the threshold level. Maintenance of constant-level voltage is difficult in case of RC-type circuit of discharge whereas in transistor-based circuits, the conditions required for self-sustenance of spark discharge is met after the breaking of the gap and is maintained till the end of the spark discharge. The above discussion clearly suggests that significant differences are observed with different sources of power supply and therefore the type of supply of power plays an important role in various applications of micro-EDM process.

Spark gap monitoring mechanisms also play a vital role in EDM applications. Sparking phenomenon of EDM can be divided into three important phases, i.e., ignition, heating, and material removal phase. Formation of embryonic plasma incepts in the ignition phase which itself is divided into two substages: formation of bubble and mechanism of electronic impact. Heating and, therefore, evaporation of dielectric liquid takes place in the bubble formation stage whereas in electronic impact phase, formation of low-density area takes place in the dielectric liquid. Localized ionization process takes place due to higher mobility of electrons which ultimately results in electron impact. Bubble formed in the ignition phase gets transformed to full plasma during the phase of heating and during the material removal stage, it disappears.

The gap monitoring mechanism in place must be able to detect differences between arcs, sparks, open, and short circuits with sampling periods of nanosecond order (Kao and Shih 2006). Research efforts have been directed to aid in judicious pulse condition monitoring mechanisms and, therefore, investigations have been carried out on acoustic techniques (Richardson and Gianchandani 2010) and fuzzy integrated high-speed data acquisition systems (Kao and Shih 2006). High-speed data acquisition systems with fuzzy controllers have reported to have suppressed unwanted arcs and, therefore, have smoothened and stabilized the micro-EDM process (Kao et al. 2008).

One of the other stringent requirements on the applicability of micro-EDM is the usage of motion platforms. Motion platforms must be able to position the target material at submicron levels. These should also be able to control the gap between the electrode feed as well as the feed rate of cutting process (Liu et al. 2010). Inter-electrode gap results in another differentiating factor between conventional EDM and micro-EDM process. Inter-electrode gap in case of micro-EDM is of the order of several microns whereas in case of conventional EDM it varies from several microns to few millimeters. Gap voltage forms the indirect measurement tool for determining the inter-electrode gap (Mahendran et al. 2010). Hydraulic and electromechanical systems are used to maintain the inter-electrode gap with positioning accuracies less than 1 $\mu$m (Mahendran et al. 2010).

Efforts have been directed by researchers to enhance the machining characteristics of the micro-EDM process. Han et al. (2004) have developed a new transistor-type impulse generator for micro-EDM machining performance improvement. The pulse duration was reduced to 30 ns using the proposed circuit. Further, the material removal rate was revealed to increase up to 24 times higher than the conventional circuits.

## 3.3  Micro-EDM Process and Its Mechanics

It is necessary to figure out the important process parameters and to optimize them for enhanced machining efficiency. However, in the case of micro-EDM, the parameters controlling the process bears a complex relationship with the mechanics of the process and therefore their optimization is a bit difficult task. Therefore, it becomes quintessential to first comprehend the mechanics of micro-EDM process and then identify the complex relationship between the process parameters.

The discussion on the mechanics of micro-EDM process can be made by dividing the micro-EDM process into three main phases: the mechanism of material removal, debris flushing, and the process parameters affecting the processes of material removal and flushing of debris.

### 3.3.1  Mechanism of Material Removal

The material removal mechanism of the conventional EDM process can be replicated to comprehend the principles of material removal in case of micro-EDM process. The material removal process in EDM involves the creation of plasma channel between the electrode and the workpiece (Katz and Tibbles 2005). As a result of the intense heat from the plasma channel, some of the material reaches their boiling point and are vaporized while the rest of the material is only heated close to their melting temperature which results in the formation of molten pool on the surface of the workpiece material (Masuzawa 2000). Collapsing of the plasma channel occurs at the end of the spark discharge and then some of the removed material gets into the bulk dielectric liquid resulting into debris (Descoeudres 2006).

Generation of spark discharge is a very rapid process with pulse on time varying from nanoseconds to microseconds. Further, the duty cycles range 60–90%. The two combinatorial results in large amount of material removal over time. Application of voltage potential across the target material and the tool electrode incepts the process of creation of spark discharge. A dielectric liquid medium acts as an insulator that prevents the flow of current across the gap. Generated high electric field results in ionization of the dielectric medium in the inter-electrode gap. This, in turn, initiates the process of dielectric breakdown. Dielectric breakdown may take place in accordance with either the basic theory linked to the growth of vapor bubble or the theory suggesting the formation of streamers between the electrodes (Dhanik and Joshi 2005).

Electronic avalanche results in streamers and these get converted to weakly ionized channels once the electronic avalanche reaches the desired amplification (Descoeudres 2006). Formation of positive streamers occurs when the distance between the workpiece and the tool electrode is small and the voltage is moderate. The streamer incepts at anode and grows toward cathode (Descoeudres 2006; Raizer 1991). However, when there is large inter-electrode gap or the voltages are high, negative streamers form that grow sufficiently in size before reaching the anode. Transition of avalanche to streamer takes place in this case and the propagation of streamer is toward both the electrodes (Descoeudres 2006; Raizer 1991).

After the completion of ionization time, breakdown of the dielectric medium takes place and therein the conversion to highly ionized plasma of the weakly ionized streamer channel takes place. Highly ionized channel of plasma is formed between the target material and the electrode tool. Heating of the target material takes place through the current passing in the plasma channel during the discharge. As a result of heating, melting and vaporization of the workpiece material take place (Yeo et al. 2007). The rate of material removal taking place due to vaporization phenomenon is very minute in comparison to that occurring due to melting (Wong et al. 2003). Melted workpiece material remains in the crater and is not eliminated until the end of the discharge.

The mechanism involving the removal of molten material is very complex involving various forces such as electrodynamics, hydrodynamics, thermodynamics, and electromagnetics (Boothroyd 2006). Research efforts have been made to develop and test different material removal models that assume that the pressure is exerted on the molten pool during the discharge by the plasma channel (Gray and Pharney 1974; Dhanik and Joshi 2005; Soldera and Mücklich 2004). The pressure is released on with the imploding plasma channel which takes place at the end of the discharge process. The process in turn results in recoil effect resulting in ejection of workpiece material from the surface of the melted workpiece material (Gray and Pharney 1974). Resolidification of debris takes place into globules as the molten material enters the dielectric liquid which may enter the inter-electrode gap.

The debris entering the gap between the electrodes are required to be flushed away from the machining zone. Trapping of debris in inter-electrode gap is a very serious problem owing to the very small inter-electrode gap. As a result of trapping, the electrical conductivity in the gap increases resulting in short circuits instead of the desired discharges. This ultimately results in reduction in efficiency of the process (Kiran and Joshi 2007). Further, the energy distribution in the plasma channel also gets altered which further reduces the efficiency of each spark energy (Kiran and Joshi 2007). Wang et al. (2009) simulated the debris movement in deep hole drilling using the micro-EDM process. It was revealed from the simulation that the debris accumulated near the electrode and workpiece surfaces. Improvement in machining stability has been reported with successful elimination of the debris (Tong et al. 2008).

### 3.3.2   Process Parameters

Open gap voltage, no-load voltage $V$, pulse current, pulse off time, and on time are the discharge pulse parameters that are controlled commonly to control the machining efficiency of the micro-EDM process. The voltage applied to the electrode before the discharge is known as no-load voltage. No-load voltage is the source of electric field that initiates the process of dielectric breakdown (Katz and Tibbles 2005). The current that flows through the plasma channel is the pulse current. The time period between the initiation of the dielectric breakdown and the implosion of plasma is the pulse on time and the time period incepting from the end of spark discharge and prior to the initiation of new discharge pulse is known as pulse off time (Kao and Shih 2006).

Energy of discharge that effects the crater dimensions is the product of voltage, pulse on time, and current. Process accuracy, electrode wear, and material removal rates are also significantly affected by discharge energy (Wong et al. 2003). Small dimensions crater are created with the small discharge energies, i.e., reduction in electrode wear and material removal rate. However, process accuracy is improved owing to the finer resolution associated with the decreased discharge crater size (Wong et al. 2003). As for the case of high discharge energy, the opposite is true. The inter-electrode gap distance is affected by no-load voltage. The increased inter-electrode gap distance calls for higher no-load voltage for initiation of the dielectric breakdown. Improvement in circulation of dielectric fluid has been reported with higher inter-electrode gap distance and also prevents the debris build up. However, poor surface finish and dimensional accuracy have been associated with the increased inter-electrode gap (Jahan et al. 2008).

Pulse off time significantly affects the debris removal process as well as the recovery process for strength of the dielectric medium (Mahendran et al. 2010). Optimum machining time and its balancing with the debris flushing process is achieved with the adjustment of pulse off time. The debris will be efficiently removed from the machining zone in case of long pulse off time. Efficient flushing results in reduction of probable chances of abnormal discharges. However, there will be a drastic reduction in the material removal rate. On the other hand, if the pulse off time is sufficiently short then the debris will remain trapped within the discharge gap and the dielectric fluid will then not be able to regain the required strength ultimately resulting in abnormal discharges (Mahendran et al. 2010). Reduced material removal rate as well as poor surface quality of the workpiece surface will, therefore, be achieved in this case (Yu et al. 2002). Therefore, it is necessary to adjust the pulse off time for balancing the debris flushing and machining time.

### 3.3.2.1 Material of Electrode

The typical electrode material includes copper and tungsten. Investigations of effect on micro-EDM process of tungsten and its variants, i.e., copper tungsten and silver tungsten have been carried by Jahan et al. (2009). The investigations have revealed that best surface finish has been obtained with silver tungsten. Highest material removal rate was obtained with copper tungsten electrode material. The wear rate obtained was lowest with tungsten electrode. Tungsten carbide was also investigated for the micro-EDM process (Liu et al. 2005; Yan et al. 2002; Weng and Her 2002; Her and Weng 2001).

#### 3.3.2.2   Polarity of Electrode

Electrode polarity can be set either as negative or positive. When the polarity of the electrode is positive, it indicates that the target material is the anode. The movement of electrons in the plasma channel is directed toward the target material. While negative polarity suggests that the anode is the electrode tool. Increased material removal rate and decreased electrode wear have been reported with the negative polarity (Jahan et al. 2008). This may be attributed to the reason that higher discharge energy is concentrated at the cathode which experiences greater heating in comparison to the anode (Kunieda et al. 2005). Further, the tool erosion is prevented due to the formation of protective carbon or oxide layer coating on employing water-based dielectrics or hydrocarbon-based dielectrics (Lin et al. 2000).

#### 3.3.2.3   Feedrate

The inter-electrode gap distance can be well maintained with the adjustment of electrode feedrate. Slower feedrate signifies a larger inter-electrode gap distance than the required optimal distance. A larger inter-electrode gap will aid in efficient flushing of the debris but at the same time signifies a slower discharge process, i.e., occurrence of fewer sparks. While on the other hand, a higher feedrate will result in too short inter-electrode gap resulting in higher occurrence of short-circuit discharges and thereby restricting the debris flushing process.

#### 3.3.2.4   Strength of Dielectric Fluid

The debris flushing is improved with the usage of dielectric fluid with weak strength. This is because the same gap voltage will be able to initialize breakdown of dielectric medium over a greater distance. The reverse is applicable for dielectric fluids with greater strengths (Wang et al. 2005). Further, the thermal conductivity as well as the viscosity of the dielectric fluid also plays a critical role in effective flushing of debris (Lin et al. 2007).

### 3.4   Working Mechanism of Magnetic Field Assisted Micro-EDM Process

Figure 3.2 depicts the schematic of magnetic field assisted micro-EDM process. Material removal process is largely dependent on the plasma channel. Plasma channel provides the required thermal energy that heats and vaporizes the workpiece material. The removal of the molten material is dependent largely on the collapse of the plasma channel. Therefore, changing the characteristics of the plasma channel will largely effect the process of material.

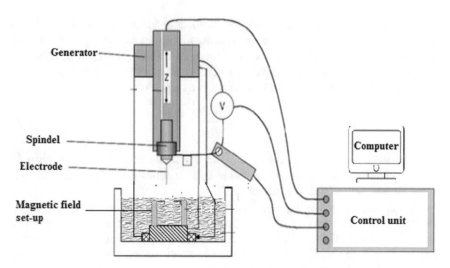

**Fig. 3.2** Schematic of magnetic field assisted micro-EDM process

Influence of the magnetic field on the plasma channel has been a topic of great interest amongst the researchers since 1950s. However, confinement and stabilization of plasma have been the major area of interest amongst the scientific community in the domain of plasma and magnetic field interaction. High-density plasmas could be obtained with the plasma confinement process besides being used for other purposes such as generation of X-rays (Haines et al. 2000). Higher current density is also achievable with the plasma confinement process which ultimately leads to increased anodic and cathodic spot heating. Sputtering process is one of the application that takes advantage of plasma confinement. Applications for plasma stabilization includes the development of plasma torches that are employed for surface processing. The research aimed at examination of plasma stabilization and confinement aids in understanding the hypothesis related to the effect magnetic field on the plasma in micro-EDM process.

The changes in plasma characteristics can be determined using the spectroscopic measurement techniques. Such techniques aids in the determination of electron density and temperature of plasma (Descoeudres 2006). Line-pair method is used for determination of the plasma temperature and is given by Eq. (3.1) (Albinski et al. 1996)

$$T = \frac{-(E_1 - E_2)/k}{\ln\left(\frac{I_1}{I_2} \frac{\lambda_1}{\lambda_2} \frac{g_2}{g_1} \frac{A_2}{A_1}\right)} \tag{3.1}$$

where $E_n$, $I_n$, $g_n$ $A_n$ and $\lambda_n$ are the energy of excitation, intensity, statistical weight, and transition probability, respectively, for the spectral line $n$, $k$ is the Boltzmann constant.

The axial magnetic field lines, i.e., the magnetic field lines that are coaxial with the electric field lines have the likely possibility for plasma confinement on the basis of Larmor radius principle. Electrons will travel in a helical path running along the field lines in case of uniform magnetic field. The radius of electron path movement known as Larmor radius is given by the following equation:

$$r = \frac{mv}{|q|B} \tag{3.2}$$

where $m$, $v$ and $q$ are the mass, velocity, and charge of electron, $B$ is the magnetic field strength. Electrons are confined in case the Larmor radius is smaller in comparison to the plasma radius (Keidar et al. 1996).

Plasma stabilization using magnetic field was established in the year 1957 by Tayler (1957), where the arc inside the gas-filled cylinder was successfully stabilized using the magnetic field. DC arc was modeled in the presence of an axial magnetic field by Kotalík and Nishiyama (2002). A more stable arc was reported owing to the decreased electronic turbulence. Suppression of recirculation zone is another reason that has been reported for the additional stability in plasma under the influence of perpendicular magnetic fields (Kotalík and Nishiyama 2002) Transverse magnetic field has been implemented by Kim (2009) and was successfully able to suppress electronic turbulence. The instabilities were found to decrease by 28.6%.

Plasma confinement as well as stabilization process will aid in lending stability to the discharge channel and thereby improve the efficiency of the micro-EDM process. Application of plasma stabilization and confinement techniques to micro-EDM process results in increased density of the current at the workpiece surface which ultimately leads to efficient material removal process.

Besides stabilizing and confining the discharge channel, effective debris flushing is also critical to improving the efficiency of micro-EDM process. There have been umpteen efforts on improvising the debris flushing process. Ultrasonic vibration (Kremer et al. 1989; Yeo and Tan 1999; Yan et al. 2002; Wansheng et al. 2002; Gao and Liu 2003), alteration of the electrode shape (Yan et al. 1999), providing planetary motion to the electrode (Yu et al. 2002), employing gravitational force (Murali and Yeo 2004) are some of the efforts made in this direction. Application of magnetic field is another means to enhance the debris flushing process and hence the process is known as magnetic-assisted micro-EDM.

Under the influence of magnetic field, the debris particle experiences two kinds of forces: magnetic and centrifugal force. Any particle that is kept inside the magnetic field can be considered as a dipole. Equation (3.3) can be used to determine the magnetic force that a dipole experiences under the influence of nonuniform magnetic field

$$\vec{F_B} = \nabla\left(\vec{\mu} \cdot \vec{B}\right) \tag{3.3}$$

where
$\nabla$ is the gradient operator, $\vec{F}_B$ denotes magnetic force, $\vec{\mu}$ is the dipole, and $\vec{B}$ indicates the nonuniform magnetic field.

Equation (3.4) is used to obtain the centrifugal force

$$\vec{F}_c = \left(\frac{mv^2}{r}\right) \tag{3.4}$$

where
$\vec{F}_c$ denotes the centrifugal force, $v$ is the velocity of debris, $m$ signifies the mass of the debris, and $r$ denotes the radial distance of the particle from the axis of electrode rotation.

Equation (3.5) is then used to obtain the resultant force on the debris

$$\vec{F} = \vec{F}_B + \vec{F}_C \tag{3.5}$$

Addition of external magnetic force, therefore, add to increased force and therefore aid in removing the inter-electrode gap trapped debris. The fact has been corroborated by the study carried out by Amson (1965). However, the presence of an external magnetic field increases the tool wear and surface roughness owing to the distortions in the tool electrode (Yeo et al. 2004).

## 3.5  Conclusion

The present chapter is an overview on one of the least explored hybrid micromachining process, i.e., magnetic field assisted micro-EDM process. It can be concluded that the debris removal process is enhanced with the presence of magnetic field. Further, the stabilization and confinement of the plasma also result in enhanced machining efficiency of the micro-EDM process. However, at the same time, some distortions are also inevitable in the tool electrode. The distortions, in turn, lead to increased tool wear of the electrode tool along its length. Surface roughness is not effected significantly in the presence of magnetic field.

## References

K. Albinski, K. Musiol, A. Miernikiewicz, S. Labuz, M. Malota, The temperature of a plasma used in electrical discharge machining. Plasma Sources Sci. Technol. 5(4), 736 (1996)
D.M. Allen, A. Lecheheb, Micro electro-discharge machining of ink jet nozzles: optimum selection of material and machining parameters. J. Mater. Process. Technol. 58(1), 53–66 (1996)
J.C. Amson, Lorentz force in the molten tip of an arc electrode. Br. J. Appl. Phys. 16(8), 1169 (1965)

G.G. Boothroyd, *Fundamentals of Machining and Machine Tools*, 3rd edn. (CRC/Taylor and Francis, Boca Raton, FL, 2006)

A. Descoeudres, *Characterization of EDM Plasmas* (Ph. D. thesis, Ecole Polytechnique Federale De Lausanne, 2006)

S. Dhanik, S.S. Joshi, Modeling of a single resistance capacitance pulse discharge in micro-electro discharge machining. J. Manuf. Sci. Eng. **127**(4), 759–767 (2005)

C. Gao, Z. Liu, A study of ultrasonically aided micro-electrical-discharge machining by the application of workpiece vibration. J. Mater. Process. Technol. **139**(1–3), 226–228 (2003)

M. Ghoreishi, J. Atkinson, A comparative experimental study of machining characteristics in vibratory, rotary and vibro-rotary electro-discharge machining. J. Mater. Process. Technol. **120**(1–3), 374–384 (2002)

E.W. Gray, J.R. Pharney, Electrode erosion by particle ejection in low-current arcs. J. Appl. Phys. **45**(2), 667–671 (1974)

M.G. Haines, S.V. Lebedev, J.P. Chittenden, F.N. Beg, S.N. Bland, A.E. Dangor, The past, present, and future of Z pinches. Phys. Plasmas **7**(5), 1672–1680 (2000)

F. Han, S. Wachi, M. Kunieda, Improvement of machining characteristics of micro-EDM using transistor type isopulse generator and servo feed control. Precis. Eng. **28**(4), 378–385 (2004)

M.G. Her, F.T. Weng, Micro-hole maching of copper using the electro-discharge machining process with a tungsten carbide electrode compared with a copper electrode. Int. J. Adv. Manuf. Technol. **17**(10), 715–719 (2001)

M.P. Jahan, Micro-electrical discharge machining, in *Nontraditional Machining Processes* (Springer, London, 2013), pp. 111–151

M.P. Jahan, Y. San Wong, M. Rahman, A comparative study of transistor and RC pulse generators for micro-EDM of tungsten carbide. Int. J. Precis. Eng. Manuf. **9**(4), 3–10 (2008)

M.P. Jahan, Y.S. Wong, M. Rahman, A study on the quality micro-hole machining of tungsten carbide by micro-EDM process using transistor and RC-type pulse generator. J. Mater. Process. Tech. **209**(4), 1706–1716 (2009)

V.K. Jain, *Introduction to Micromachining* (Alpha Science International Limited, Oxford, 2010)

C.C. Kao, A.J. Shih, Sub-nanosecond monitoring of micro-hole electrical discharge machining pulses and modeling of discharge ringing. Int. J. Mach. Tools Manuf. **46**(15), 1996–2008 (2006)

C.C. Kao, A.J. Shih, S.F. Miller, Fuzzy logic control of microhole electrical discharge machining. J. Manuf. Sci. Eng. **130**(6), 064502 (2008)

Z. Katz, C.J. Tibbles, Analysis of micro-scale EDM process. Int. J. Adv. Manuf. Technol. **25**(9–10), 923–928 (2005)

M. Keidar, I. Beilis, R.L. Boxman, S. Goldsmith, 2D expansion of the low-density interelectrode vacuum arc plasma jet in an axial magnetic field. J. Phys. D Appl. Phys. **29**(7), 1973 (1996)

K.S. Kim, Influence of a transverse magnetic field on arc root movements in a dc plasma torch: diamagnetic effect of arc column. Appl. Phys. Lett. **94**(12), 121501 (2009)

M.K. Kiran, S.S. Joshi, Modeling of surface roughness and the role of debris in micro-EDM. J. Manuf. Sci. Eng. **129**(2), 265–273 (2007)

P. Kotalík, H. Nishiyama, An effect of magnetic field on arc plasma flow. IEEE Trans. Plasma Sci. **30**(1), 160–161 (2002)

D. Kremer, J.L. Lebrun, B. Hosari, A. Moisan, Effects of ultrasonic vibrations on the performances in EDM. CIRP Ann. Manuf. Technol. **38**(1), 199–202 (1989)

M. Kunieda, B. Lauwers, K.P. Rajurkar, B.M. Schumacher, Advancing EDM through fundamental insight into the process. CIRP Ann. Manuf. Technol. **54**(2), 64–87 (2005)

B. Lauwers, J.P. Kruth, W. Liu, W. Eeraerts, B. Schacht, P. Bleys, Investigation of material removal mechanisms in EDM of composite ceramic materials. J. Mater. Process. Technol. **149**(1–3), 347–352 (2004)

Y.C. Lin, B.H. Yan, Y.S. Chang, Machining characteristics of titanium alloy (Ti–6Al–4V) using a combination process of EDM with USM. J. Mater. Process. Technol. **104**(3), 171–177 (2000)

C.T. Lin, H.M. Chow, L.D. Yang, Y.F. Chen, Feasibility study of micro-slit EDM machining using pure water. Int. J. Adv. Manuf. Technol. **34**(1–2), 104–110 (2007)

K. Liu, B. Lauwers, D. Reynaerts, Process capabilities of Micro-EDM and its applications. Int. J. Adv. Manuf. Technol. **47**(1-4), 11-19 (2010)

H.S. Liu, B.H. Yan, F.Y. Huang, K.H. Qiu, A study on the characterization of high nickel alloy micro-holes using micro-EDM and their applications. J. Mater. Process. Technol. **169**(3), 418–426 (2005)

S. Mahendran, R. Devarajan, T. Nagarajan, A. Majdi, A review of micro-EDM, in *Proceedings of the International Multi Conference of Engineers and Computer Scientists*, vol. 2

T. Masuzawa, State of the art of micromachining. CIRP Ann. Manuf. Technol. **49**(2), 473–488 (2000)

M. Murali, S.H. Yeo, A novel spark erosion technique for the fabrication of high aspect ratio micro-grooves. Microsyst. Technol. **10**(8–9), 628–632 (2004)

A. Muttamara, Y. Fukuzawa, N. Mohri, T. Tani, Probability of precision micro-machining of insulating $Si_3N_4$ ceramics by EDM. J. Mater. Process. Technol. **140**(1–3), 243–247 (2003)

Y.P. Raizer, *Gas Discharge Physics* (1991)

K.P. Rajurkar, G. Levy, A. Malshe, M.M. Sundaram, J. McGeough, X. Hu, R. Resnick, A. DeSilva, Micro and nano machining by electro-physical and chemical processes. CIRP Ann. Manuf. Technol. **55**(2), 643–666 (2006)

M.T. Richardson, Y.B. Gianchandani, Wireless monitoring of workpiece material transitions and debris accumulation in micro-electro-discharge machining. J. Microelectromech. Syst. **19**(1), 48–54 (2010)

B. Shao, K.P. Rajurkar, Modelling of the crater formation in micro-EDM. Proc. CIRP **33**, 376–381 (2015)

F.A. Soldera, F. Mücklich, on the erosion of material surfaces caused by electrical plasma discharging, in *MRS Online Proceedings Library Archive* (2004), p. 843

X. Song, W. Meeusen, D. Reynaerts, H. Van Brussel, Experimental study of micro-EDM machining performances on silicon wafer, in *Micromachining and Microfabrication Process Technology VI*, vol. 4174 (International Society for Optics and Photonics, 2000), pp. 331–340

R.J. Tayler, The influence of an axial magnetic field on the stability of a constricted gas discharge. Proc. Phys. Soc. Sect. B **70**(11), 1049 (1957)

H. Tong, Y. Li, Y. Wang, Experimental research on vibration assisted EDM of micro-structures with non-circular cross-section. J. Mater. Process. Technol. **208**(1–3), 289–298 (2008)

A.C. Wang, B.H. Yan, Y.X. Tang, F.Y. Huang, The feasibility study on a fabricated micro slit die using micro EDM. Int. J. Adv. Manuf. Technol. **25**(1–2), 10–16 (2005)

J. Wang, Y.G. Wang, F.L. Zhao, Simulation of debris movement in micro electrical discharge machining of deep holes, in *Materials Science Forum*, vol. 626 (Trans Tech Publications, 2009), pp. 267–272

Z. Wansheng, W. Zhenlong, D. Shichun, C. Guanxin, W. Hongyu, Ultrasonic and electric discharge machining to deep and small hole on titanium alloy. J. Mater. Process. Technol. **120**(1–3), 101–106 (2002)

F.T. Weng, M.G. Her, Study of the batch production of micro parts using the EDM process. Int. J. Adv. Manuf. Technol. **19**(4), 266–270 (2002)

Y.S. Wong, M. Rahman, H.S. Lim, H. Han, N. Ravi, Investigation of micro-EDM material removal characteristics using single RC-pulse discharges. J. Mater. Process. Technol. **140**(1–3), 303–307 (2003)

B.H. Yan, F.Y. Huang, H.M. Chow, J.Y. Tsai, Micro-hole machining of carbide by electric discharge machining. J. Mater. Process. Technol. **87**(1–3), 139–145 (1999)

B.H. Yan, A.C. Wang, C.Y. Huang, F.Y. Huang, Study of precision micro-holes in borosilicate glass using micro EDM combined with micro ultrasonic vibration machining. Int. J. Mach. Tools Manuf. **42**(10), 1105–1112 (2002)

S.H. Yeo, L.K. Tan, Effects of ultrasonic vibrations in micro electro-discharge machining of micro-holes. J. Micromech. Microeng. **9**(4), 345 (1999)

S.H. Yeo, M. Murali, H.T. Cheah, Magnetic field assisted micro electro-discharge machining. J. Micromech. Microeng. **14**(11), 1526 (2004)

S.H. Yeo, W. Kurnia, P.C. Tan, Electro-thermal modelling of anode and cathode in micro-EDM. J. Phys. D Appl. Phys. **40**(8), 2513 (2007)

Z.Y. Yu, K.P. Rajurkar, H. Shen, High aspect ratio and complex shaped blind micro holes by micro EDM. CIRP Ann. Manuf. Technol. **51**(1), 359–362 (2002)

# Chapter 4
# Electrorheological Fluid-Assisted Micro-USM

## 4.1 Introduction

Miniaturization of products has occupied the center stage in recent years. Publications in the field of micromachining started in the last decade and since then it has gained momentum owing to the wide acceptability meeting the industrial needs. There has been a strong inclination for miniaturized products because of number of associated advantages such as space requirement, consumption of lesser energy, lesser amount of material consumption, and flexibility in handling. However, there is umpteen number of challenges such as innovation and continuous process improvements in manufacturing technologies to process a wide variety of engineering materials (Masuzawa 2000). Research is therefore grown exponentially to invent new manufacturing technologies as well as to improvise the existing techniques to achieve higher dimensional precisions and economic cost of production. An array of issues associated with the micromachining technologies have been reported by researchers (Gentili et al. 2007).

Few of the bulk micromachining techniques, the concept of which was developed in 1983, have been developed and have been employed for producing pressure sensors, ink nozzles, and accelerometers (Allen 2005). Although the bulk micromachining techniques have been used for mass production of microchip technology, but their industrial application has been limited by geometrical limits and the higher cost of production. The devices produced are relatively large and therefore occupy most of the area on chip (French et al. 1997).

3D free-form micro-features have been generated using tool-based micromachining methods. Few instances of tool-based micromachining methods are microgrinding, micro-turning, and micromilling. However, restrictions on microcutting processes have been restricted because of low stiffness of the cutting tool and also by its low strength. The restrictions are applicable especially for hard materials (Fang et al. 2006; Liow 2009). Aforementioned typical restrictions can, however,

© The Author(s), under exclusive license to Springer Nature Switzerland AG 2019
S. Bhowmik and D. Zindani, *Hybrid Micro-Machining Processes*,
SpringerBriefs in Applied Sciences and Technology,
https://doi.org/10.1007/978-3-030-13039-8_4

be overcome by employing nontraditional machining technologies such as micro-discharge micromachining, laser micromachining, micro-ultrasonic machining, and micro-electrochemical machining.

Various nontraditional micromachining processes have umpteen number of associated advantages but are plagued by certain issues such as the creation of heat affected zone (HAZ) and the thermal stresses such as that in the case of micro-EDM and laser micromachining processes (Singh et al. 2008; Hung et al. 2006), the decreased rate of dissolution with the increasing depth of machining in case of electrochemical micromachining (Yang et al. 2009) and so on. Ultrasonic vibrations have been used to improvise the machining process such as that in the case of micro-EDM to aid in efficient removal of the debris (Endo et al. 2008; Koshimizu and Iansaki 1988). The assistance of vibrations to ultrasonic machining giving rise to micro-ultrasonic machining has also been explored by the scientific community (Jain et al. 2011).

Ion etching is yet another least explored technique that can be used in machining of hard brittle material at higher precision levels. The technique, however, demands a customized environment and equipment. The aim of the present chapter is to brief the readers with the electrorheological fluid-assisted micro-USM micromachining process. The chapter begins with a brief outline on micro-USM process and then advances to discuss on the mechanism behind the generation of chippings. Next, the mechanism of electrorheological fluid-assisted micro-USM micromachining process has been discussed. The chapter ends with conclusive remarks.

## 4.2   Overview of Micro-USM Process

Ultrasonic machining process incepted in the year 1927. It was patented by L. Balamuth in the year 1945. A number of connotations have existed since then for the micromachining process such as impact grinding, slurry drilling, etc. Figure 4.1 depicts the schematic of the micro-USM process. The material removal mechanism depends on the cutting action of the abrasive slurry that flows between the workpiece and the transducer tip. The first USM tool was built in the year 1953 and by 1960s commercialisation of the various USM tools took place in bulk (Kremer et al. 1981). Since its inception, ultrasonic machining operation has gained ground in manufacturing field and its industrial application is growing exponentially (Thoe et al. 1998).

Micro-USM stems out from the general ultrasonic machining process. It was during 1990s that the efforts were made by Masuzawa of Tokyo University to downsize the macro-USM process to microlevel (Sun et al. 1996). The micro-USM process has the capabilities of producing surfaces free from thermal stresses and machining of nonconductive and hard materials that have brittle characteristics. Micro-USM process has been investigated for machining of typical materials such as silicon, alumina and glass (Ghahramani and Wang 2001; Masuzawa and Tönshoff 1997).

**Fig. 4.1** Process diagram of micro-USM

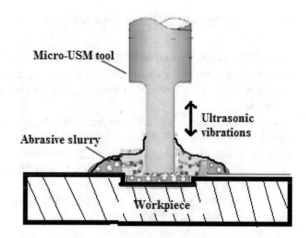

The process parameters of micro-USM are similar to that of macro-USM process. Micro-USM process, however, requires the employability of micro-sized tool and abrasive particles. Much smaller amplitudes are also another major requirement for micro-USM process.

The basic setup of micro-USM process comprises of slurry supply unit and a tool system. Mechanical vibrations with frequency ranging 20–40 kHz and amplitudes within few millimeters are employed by the micro-USM process. Abrasive slurry is a mixture of micro-sized abrasive grain particles (usually up to 5 μm) and a liquid medium. This abrasive slurry mixture is fed into the gap between the workpiece material and the tool. The fine abrasive particles gain momentum once hit by the vibrating tool head and then they impact the target workpiece material. Continued impact leads to the generation of fatigue stress localized in the impact zone. Generation of fatigue stress thereby results in the material removal mechanism. A small amount of material removal also takes place by the mechanical abrasion action of the micro-abrasive grains. Cavitation phenomenon created on account of imploding bubble also results in the removal of material at microlevel. The liquid medium making up the contents of abrasive slurry is usually water. Care should be taken to keep the abrasive slurry free from any chemical impurities since their presence can harm the workpiece material surface. The debris produced during the process are washed away by the flowing abrasive slurry and the gap is refilled with fresh slurry.

The process has been widely used for machining of micro-holes. However, creation of chippings around the edge of micro-holes is one of the major drawbacks in micro-USM process. The chippings become larger when the machining depth becomes larger. It is necessary to prevent the formation of chippings in order to achieve machining with high precision. The subsequent section briefs on the mechanism of chipping generation.

## 4.3   Chippings and Their Generation

Generation of chippings is one of the major drawback associated with the machining of hard and brittle materials using USM. Tool size does not effect the size of chippings. In order to achieve high-precision machining, it is essential to prevent the generation of chippings.

Generation of chippings with the increasing depth of the micro-hole was investigated by Tateishi et al. (2008). No chippings were observed till the machining depth of 5 $\mu$m. Generation of chippings incepted when the depth of machining increased to 11 $\mu$m. When the machining depth was more than 25 $\mu$m, chippings were revealed to be observed all around the edge of the micro-hole. The study, therefore, revealed that the generation of chippings starts in the early stage of machining of micro-hole i.e., at a very small machining depth.

Further, the abrasive grain behavior in the vicinity of the machining zone, revealing the reason for the occurrence of chippings was also investigated in the same study. For making such observations a mirror is attached to the micro-USM stage and is tilted at certain angle. A CCD camera is then used to observe the machining area straight downward through the mirror. The data observed is saved on a computer. The behavior of the abrasive grains could be clearly observed with the larger sized abrasive particles. Abrasive grains were observed to be present before the addition of ultrasonic vibrations to the tool. However, the abrasive particles were found to be eliminated under the influence of ultrasonic vibrations.

The observations clearly indicate that the material removal mechanism not only takes place with the impacting abrasive particles but also by the direct tool tip contact with the target material. Therefore it can be concluded that the generation of chippings can be attributed to the direct contact of the tool tip. This fact has also been corroborated with the machining of micro-holes without the usage of abrasive grains. A lot of chippings around the edge of the micro-hole was observed. The absence of abrasive grains in the machining area and the direct contact of the tool are the major reasons behind the generation of chippings. Therefore it is quintessential to prevent the flushing of abrasive particles from the machining area in order to reduce the chances of chipping generation.

## 4.4   Electrorheological Fluid-Assisted Micro-USM

One of the solutions to restrain the generation of chippings is through the usage of electrorheological fluid (ER fluid). It has been revealed by the past researchers that with the increasing intensity of the electric field, viscosity of ER fluid increases (Kaku et al. 2006; Kuriyagawa et al. 2002). Si-oil and dielectric particles form the ingredients of an ER fluid. In the absence of electric field, the dielectric particles remain dispersed in Si-oil. On application of the electric field, polarization of the dielectric

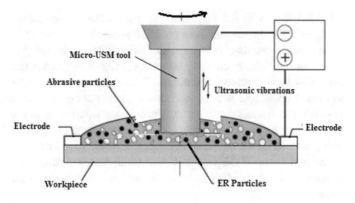

**Fig. 4.2**  Schematic of electrorheological fluid-assisted micro-USM

particles take place and align themselves in chain-like structure. The attractive force between the dielectric particles result in increased viscosity of the ER fluid.

In the electrorheological fluid-assisted micro-USM process, depicted in Fig. 4.2, the electric field is applied to the machining area by placing electrode on the work-piece. The gap between the tool and the workpiece material is filled using the mixture of abrasive grains and ER fluid. An electric force F is experienced by an object that is under the influence of nonuniform electric field. Even the electro-neutral objects experience electric force under the influence of nonuniform electric field $E$. Electric force $F$ is calculated using Eq. (4.1) (Linoya 1989). The objects in the machining area have been assumed to be spherical with diameter $d_p$.

$$F = \frac{1}{4}\pi d_p^3 \varepsilon \left( \frac{\varepsilon_p - \varepsilon}{\varepsilon_p + 2\varepsilon} \right) \text{grad} |E|^2 \tag{4.1}$$

where $\varepsilon_p$ and $\varepsilon$ are dielectric constants of the object and ER fluid.

On experiencing the electric force the abrasive particles move towards the tip of the tool and the phenomenon is termed as dielectrophoresis phenomenon. How-ever, the formation of bubbles takes place owing to the cavitation phenomenon in the machining area. The formed bubbles result in suppression of abrasive particles (Kuriyagawa 2001; Tanaka et al. 2005). But, the dielectric strength of ER fluid is greater than that of the generated bubbles and therefore the electric force $F$ on the generated bubbles will act in a direction opposite to that of the ER fluid. As a result, the generated bubbles will be removed from the machining zone. The ER fluid par-ticles will align themselves in the form of a chain and the abrasive particles will concentrate themselves around the tool tip owing to the electric force $F$. Thus, the abrasive particles are not eliminated and therefore chippings can be minimized.

Investigation with ER fluid-assisted micro-USM was carried out by Tateishi et al. (2009) and it was revealed that the average area of chipping was found out to be 0.0006 mm$^3$ in compared to 0.0020 mm$^3$ achieved using conventional USM. Thus a clear decrease of 70% in the chipping area was attained with the ER fluid-assisted micro-USM. Therefore with ER fluid-assisted micro-USM, precise machining of micro-holes on hard and brittle materials could be achieved.

The spindle speed exceeding 30 rev/min has lesser effect on chippings at the hole. Effect of speed of the spindle on material removal rate was investigated by Lian et al. (2016) for both ER fluid-assisted micro-USM and for conventional USM. For conventional USM it was revealed that the material removal rate decreases with the increasing spindle speed when the spindle speed is increased from 30 to 110 rev/min. Material removal rate decreased when the spindle speed ranged between 60 and 90 rev/min and then again increased for spindle speed beyond 90 rev/min. However, in case of ER-assisted micro-USM process the material removal rate decreased with spindle speed increasing from 30 to 110 rev/min but then increases for spindle speed over and above 110 rev/min.

Lian et al. (2016) also studied the effect of voltage on chippings at the hole edge and revealed that no chippings were observed as the voltage increased from 20 to 60 V. Adequate presence of abrasive grains could be ensured in the machining area when the voltage is kept over 20 V. Therefore the generation of chippings is not effected by the voltage when it is over and above 20 V. The impact on the material removal rate of the voltage reveals that higher number of abrasive grains are drawn in the machining zone with the higher voltage. As a result, there is an increase in the material removal rate with the voltage. However, as the voltage goes beyond a certain value, the material removal rate decreases as greater number of smaller abrasive grains are drawn into the machining zone.

## 4.5   Other Fluid-Assisted Micromachining Techniques

In this section, application of water, chemicals and gas on the target material for enhanced material removal and machining performance has been discussed. A brief overview on the chemical-assisted micromachining, water-assisted micromachining, and gas-assisted micromachining has been provided.

### 4.5.1   Chemical-Assisted Micromachining

Usage of methanol in laser micro-drilling of micro-holes results in a smoother and clean surface of the generated micro-holes. Better wettability, as well as lower boiling temperature of the solvent, may be attributed as the reasons behind attainment of smoother and cleaner micro-holes. This is because these characteristics result in enhanced rate of cooling and therefore improved drilling quality. Further, the optical properties of the target surface are also modified by the thin-layer of liquid film. The absorptivity of the target surface increases with the application of liquid film owing to greater reflective index of liquid in comparison to that of the air. A regime of greater radiant exposure leads to the generation of plasma confinement in a liquid environment and therefore plasma with higher pressure, temperature and density. The plasma with aforementioned characteristics then ultimately results in explosive rate of material removal. Enhanced rate of laser ablation with improved surface quality has been revealed to be achieved with the application of methanol layer in case of chemical-assisted laser micro-drilling of SiC (Wee et al. 2011a, b). Improved material removal rate with enhanced surface quality has also been attained with the usage of salt solution in laser micromachining. Size of heat affected zone and recast was observed to be reduced in case of laser drilling as well as laser milling of stainless steel employing salt solution. Increased material removal rate has also been achieved with the employability of liquids than in comparison to air (Li and Achara 2004).

Conventional USM suffers from some of the major disadvantages such as low rate of material removal and poor quality of surface. Utilization of hydrofluoric acid having low concentration with the abrasive grains has resulted in the elimination of the aforementioned disadvantages associated with the USM process. The efficiency of the USM process was also revealed to be improved. A drastic reduction in the machining load was achieved for the micro- and macro-drilling processes with chemical-assisted USM process. However, the process resulted in an enlarged size of the machining hole and therefore low concentration of hydrofluoric acid has been recommended to be used (Choi et al. 2007).

Alteration in mechanical properties of the target material has been achieved with the usage of electrochemical passivation in electrochemical-assisted micromachining process. Lowering of the mechanical strength of the substrate surface aids in the easy material removal process. The electrochemical passivation process generates a thin oxide layer on the surface of the target material that lowers the mechanical strength of the surface. The depth of cut in the process is required to be consistent with the thickness of the formed oxide layer. As for instance, the cutting depth for electrochemical-assisted micro-turning of stainless steel was recommended to be in the range of 100 nm in a study conducted by Sebastian et al. (2014).

### 4.5.2   Gas-Assisted Variant of Hybrid Micromachining

Trepan drilling or percussion drilling is required to be done for laser micro-drilling of cooling holes in some of the components of aero-engine. This is performed at acute angles to the surface. Nimonic 263 superalloys form the major construction material of these parts. Since these components function under extreme temperature conditions, thermal barrier coatings of plasma sprayed ceramic material are provided over the components. One of the major problems in drilled angled holes is that of delamination of the thermal barrier coating. Application of the secondary gas jet is one of the options to eliminate the problem of delamination. This aids in controlling the trajectories of the melt flow and also the angle of impact on the hole walls. This is accomplished by employing an off-axis gas jet along with the laser micro-drilling process and allowing it to impinge near the leading edge of the angled hole. The angle of offset of the secondary gas jet is 90° to the axis of the beam. For coaxial jet oxygen is used whereas for the off-axis jets nitrogen is employed using converging nozzle with diameter of up to 1.5 mm. Melt is ejected with much lesser erosive characteristics at the entrance of the inclined hole and thereby preventing the problem of thermal barrier coating delamination. However, in case of micro-drilling of Nimonic substrate, the formation of cracks at the barrier coating and substrate interface was observed. This was attributed to the changes in the melt and vapor flow. But such cracks presented a lesser risk of being propagated further since these were very small both in length and width (Sezer et al. 2009).

### 4.5.3  Water-Assisted Micromachining

Water-assisted $CO_2$ and water-assisted femtosecond laser pulse ablation have been explored for ablation of micro-meter range holes with very high aspect ratio. The processes have been successfully applied in alumina, LCD glass, and silicon. Efficient drilling of holes has been achieved with the application of thin water layer on the processed surface. Less taper, heat affected zone, and micro-cracks have been revealed with the water layer in comparison to the atmospheric air. Removal of debris from the machining area is the primary reason behind the efficient drilling process. The debris produced during the ablation process blocks the transformation of energy in absence of thin water spray. As a result, the incoming laser pulses are absorbed and scattered and thereby leading to the inefficient drilling process. The presence of water results in continual removal of debris from the drilling zone and therefore the process of drilling is efficient. The shape of the hole produced is also inflicted owing to the scattering by the debris (Kaakkunen et al. 2011; Wee et al. 2011a, b; Tsai and Li 2009; Hwang et al. 2004). However, a rough surface is the major drawback of the process which is due to the rapid solidification of the molten metal (Choo et al. 2004).

Improved surface quality has been observed with the usage of water in case of laser ablation of Si and $Al_2O_3$. Enhanced ablation of material due to laser is achieved at lower fluence of laser by water for polymethyl methacrylate (PMMA). However, for polyethylene terephthalate no enhancement was revealed with the employability of water. Further smoother surface was obtained with excimer laser of PET than compared to PMMA (Jang and Kim 2006).

### 4.6  Conclusion

In the present chapter, a brief overview of the ER-assisted micro-USM process has been provided. The usage of ER fluid eliminates the formation of edge chippings in the fabrication of micro-holes using the conventional micro-USM process. With the employability of suitable processing parameters the ER-assisted micro-USM process has an obvious advantage over the conventional micro-USM process. No significant effects on the edge chippings of the voltage and ultrasonic power are observed in the case of ER-assisted micro-USM process. The material removal rate increase with the ultrasonic power in ER-assisted micro-USM process. Material removal rate in case of ER-assisted micro-USM process is higher than the conventional micro-USM process. The chapter also provides a brief outline of other fluid-assisted micromachining process highlighting the importance of fluid introduction in the enhancement of machining efficiency.

# References

J.J. Allen, *Micro Electro Mechanical System Design* (CRC Press, 2005)

J.P. Choi, B.H. Jeon, B.H. Kim, Chemical-assisted ultrasonic machining of glass. J. Mater. Process. Technol. **191**(1–3), 153–156 (2007)

K.L. Choo, Y. Ogawa, G. Kanbargi, V. Otra, L.M. Raff, R. Komanduri, Micromachining of silicon by short-pulse laser ablation in air and under water. Mater. Sci. Eng., A **372**(1–2), 145–162 (2004)

T. Endo, T. Tsujimoto, K. Mitsui, Study of vibration-assisted micro-EDM—the effect of vibration on machining time and stability of discharge. Precis. Eng. **32**(4), 269–277 (2008)

F.Z. Fang, K. Liu, T.R. Kurfess, G.C. Lim, Tool-based micro machining and applications in MEMS, in *MEMS/NEMS* (Springer, Boston, 2006), pp. 678–740

P.J. French, P.T.J. Gennissen, P.M. Sarro, New silicon micromachining techniques for microsystems. Sens. Actuators, A **62**(1–3), 652–662 (1997)

E. Gentili, L. Tabaglio, F. Aggogeri, Review on micromachining techniques, in *Proceedings of the 7th International Conference on Advance Manufacturing Systems and Technology (AMST '05)* (2007), p. 486

B. Ghahramani, Z.Y. Wang, Precision ultrasonic machining process: a case study of stress analysis of ceramic ($Al_2O_3$). Int. J. Mach. Tools Manuf. **41**(8), 1189–1208 (2001)

J.C. Hung, B.H. Yan, H.S. Liu, H.M. Chow, Micro-hole machining using micro-EDM combined with electropolishing. J. Micromech. Microeng. **16**(8), 1480 (2006)

D.J. Hwang, T.Y. Choi, C.P. Grigoropoulos, Liquid-assisted femtosecond laser drilling of straight and three-dimensional microchannels in glass. Appl. Phys. A **79**(3), 605–612 (2004)

V. Jain, A.K. Sharma, P. Kumar, *Recent Developments and Research Issues in Microultrasonic Machining* (ISRN Mechanical Engineering, 2011)

D. Jang, D. Kim, Liquid-assisted excimer laser micromaching for ablation enhancement and debris reduction. J. Laser Micro/Nanoeng. **1**(3), 221–225 (2006)

J.J.J. Kaakkunen, M. Silvennoinen, K. Paivasaari, P. Vahimaa, Water-assisted femtosecond laser pulse ablation of high aspect ratio holes. Phys. Proc. **12**, 89–93 (2011)

T. Kaku, T. Kuriyagawa, N. Yoshihara, Electrorheological fluid-assisted polishing of WC micro aspherical glass moulding dies. Int. J. Manuf. Technol. Manage. **9**(1–2), 109–119 (2006)

S. Koshimizu, I. Iansaki, Hybrid machining of hard and brittle materials. J. Mech. Work. Technol. **17**, 333–341 (1988)

D. Kremer, S.M. Saleh, S.R. Ghabrial, A. Moisan, The state of the art of ultrasonic machining. CIRP Ann. Manuf. Technol. **30**(1), 107–110 (1981)

T. Kuriyagawa, Generation and countermeasure of cavitation in tool-workpiece interface during ultrasonic machining (Studies on mechanism of ultrasonic machining, 3rd report). J. Soc. Grind. Eng. **45**(9), 442–447 (2001)

T. Kuriyagawa, M. Saeki, K. Syoji, Electrorheological fluid-assisted ultra-precision polishing for small three-dimensional parts. Precis. Eng. **26**(4), 370–380 (2002)

L. Li, C. Achara, Chemical assisted laser machining for the minimisation of recast and heat affected zone. CIRP Ann. Manuf. Technol. **53**(1), 175–178 (2004)

H.S. Lian, Z.N. Guo, J.W. Liu, Z.G. Huang, J.F. He, Experimental study of electrophoretically assisted micro-ultrasonic machining. Int. J. Adv. Manuf. Technol. **85**(9–12), 2115–2124 (2016)

K. Linoya, *Powder Technology Handbook* (Nikkan Kogyo Shimbun, 1989), pp. 311–312 (in Japanese)

J.L. Liow, Mechanical micromachining: a sustainable micro-device manufacturing approach? J. Clean. Prod. **17**(7), 662–667 (2009)

T. Masuzawa, State of the art of micromachining. CIRP Ann. Manuf. Technol. **49**(2), 473–488 (2000)

T. Masuzawa, H.K. Tönshoff, Three-dimensional micromachining by machine tools. CIRP Ann. Manuf. Technol. **46**(2), 621–628 (1997)

S. Sebastian, G. March, S. Maciej, Experimental research on electrochemically assisted microturning process, in *Key Engineering Materials* (2014)

H.K. Sezer, L. Li, S. Leigh, Twin gas jet-assisted laser drilling through thermal barrier-coated nickel alloy substrates. Int. J. Mach. Tools Manuf. **49**(14), 1126–1135 (2009)

R. Singh, M.J. Alberts, S.N. Melkote, Characterization and prediction of the heat-affected zone in a laser-assisted mechanical micromachining process. Int. J. Mach. Tools Manuf. **48**(9), 994–1004 (2008)

X.Q. Sun, T. Masuzawa, M. Fujino, Micro ultrasonic machining and self-aligned multilayer machining/assembly technologies for 3D micromachines. In *Proceedings of the IEEE, The Ninth Annual International Workshop on Micro Electro Mechanical Systems, 1996, MEMS'96. An Investigation of Micro Structures, Sensors, Actuators, Machines and Systems* (IEEE, 1996), pp. 312–317

S. Tanaka, J.I. Takagi, T. Yokosawa, N. Hasegawa, Study on ultrasonic machining of small diameter holes. J. Jpn. Soc. Abras. Technol. **49**(5), 245–249 (2005)

T. Tateishi, K. Shimada, N. Yoshihara, J.W. Yan, T. Kuriyagawa, Effect of electrorheological fluid assistance on micro ultrasonic machining, in *Advanced Materials Research*, vol. 69 (Trans Tech Publications. 2009), pp. 148–152

T. Tateishi, N. Yoshihara, J. Yan, T. Kuriyagawa, Study on electrorheological fluid-assisted microultrasonic machining. Int. J. Abras. Technol. **2**(1), 70–82 (2008)

T.B. Thoe, D.K. Aspinwall, M.L.H. Wise, Review on ultrasonic machining. Int. J. Mach. Tools Manuf. **38**(4), 239–255 (1998)

C.H. Tsai, C.C. Li, Investigation of underwater laser drilling for brittle substrates. J. Mater. Process. Technol. **209**(6), 2838–2846 (2009)

L.M. Wee, E.Y.K. Ng, A.H. Prathama, H. Zheng, Micro-machining of silicon wafer in air and under water. Opt. Laser Technol. **43**(1), 62–71 (2011a)

L.M. Wee, L.E. Khoong, C.W. Tan, G.C. Lim, Solvent-assisted laser drilling of silicon carbide. Int. J. Appl. Ceram. Technol. **8**(6), 1263–1276 (2011b)

I. Yang, M.S. Park, C.N. Chu, Micro ECM with ultrasonic vibrations using a semi-cylindrical tool. Int. J. Precis. Eng. Manuf. **10**(2), 5–10 (2009)

# Chapter 5
# Other Assisted Hybrid Micromachining Processes

## 5.1 Introduction

Micromachining processes have been accepted and widely used in wide variety of engineering domains such as information technology (Xue et al. 2015), micro-fluidic systems, biomedical, automotive, micro-electromechanical products (Ghoshal and Bhattacharyya 2015), and aerospace. With the continued efforts of the research community in the world of micromachining, efficient and better micromachining processes have been developed. Examples of such processes include micro-milling (Rodrigues and Labarga 2014), laser beam micromachining (Zeng et al. 2012), micro-electrochemical discharge machining (Shin et al. 2011), micro-electrochemical machining (Koyano and Kunieda 2013), and micro-electro discharge machining (Nguyen et al. 2013).

However, abovementioned micromachining processes have certain drawbacks associated with them. Some of the drawbacks include that of relatively longer penetration time, higher cost of machining setup, and flexibility of micromachining process. Micro-milling process, for instance, suffers from the deformation and vibration between the microtool and the target material. The machining accuracy suffers from such a drawback (Huo 2013; Rahman et al. 2001; Suzuki et al. 2007; Uhlmann et al. 2005; Friedrich et al. 1997).

Aforementioned drawbacks have been tried to resolve with continuous efforts of scientific community and as a result of such efforts a number of breakthroughs have been achieved. One such breakthrough involves the usage of vibrations to the different standalone micromachining framework. Utilization of vibrations has revealed to have better surface quality and improved tool life. The material removal process too has been reported to improve with the employability of high frequency and small-amplitude vibrations to either the tool or the workpiece or the work fluid. Other reported advantages of vibration-assisted micromachining process include reduction in micromachining forces (Zhou et al. 2003; Amini et al. 2017; Li et al. 2017; Weremczuk et al. 2015), form accuracy (Amini et al. 2017; Moriwaki and

© The Author(s), under exclusive license to Springer Nature Switzerland AG 2019    49
S. Bhowmik and D. Zindani, *Hybrid Micro-Machining Processes*,
SpringerBriefs in Applied Sciences and Technology,
https://doi.org/10.1007/978-3-030-13039-8_5

Shamoto 1991; Shamoto et al. 2002; Xiao et al. 2003) and reduction in formation of burr (Chen et al. 2018; Moriwaki and Shamoto 1991).

Another assisted micromachining process has been evolved with the assistance of electric field to the standalone micromachining process. The process is known as external electric field assisted micromachining process and has been known for its effective debris flushing mechanism. This also resulted in inhibition of the material being reposited on the target surface (Zheng and Jiang 2009). Inclusion of carbon nanofibers has been explored for the micro-EDM process. This has resulted in reduced spark gap and hence enhanced material removal rate.

Looking at the importance of the least explored assisted hybrid micromachining processes, the present chapter outlines and discusses such processes. Present chapter makes discussion on vibration-assisted micromachining, electrical field assisted micromachining process, and carbon fiber assisted micromachining process.

## 5.2   Vibration-Assisted Micromachining

In vibration-assisted micromachining processes, the standalone micromachining process is combined with the small-amplitude tool or the vibrations of target material or working fluid. The vibrations have been used for a number of standalone micromachining processes such as micro-milling, micro-EDM, etc. A number of advantages have been revealed with the vibration-assisted micromachining process such as improved tool life and surface finish as in case of vibration-assisted micro-cutting (Brehl and Dow 2008). However, surface cracks are imminent owing to the hammering action of the tool (Lauwers 2011).

Vibrations have also been used for non-cutting processes to enhance the machining performance. As for instance, increased flushing efficiency and higher material removal rate has been achieved with ultrasonic-assisted micro-electro discharge machining process. Smooth surfaces can be achieved with suitable selection of process parameters. Use of vibrations has resulted in formation of small heat-affected zone (HAZ) and hence low resulting thermal stresses (Lauwers 2011). The machining rate is enhanced with the work fluid vibration that aids in proper circulation of debris. With proper circulation of the debris, surface finish and accuracy of the standalone micromachining processes such as micro-EDM and micro-electrochemical machining process is improved.

### 5.2.1   Tool Vibration Assisted Variant of Hybrid Micromachining

Tool vibrations are generated with the usage of piezoelectric transducer, horn and a booster in tool vibration assisted micromachining processes. The piezoelectric effect results in conversion of electric impulse to mechanical vibrations that has ultrasonic frequency. Booster and horn then amplifies the amplitude of tool vibration (Tawakoli et al. 2009; Azarhoushang and Akbari 2007). Researchers have conducted study on the effects of vibrations on the surface quality, tool wear, material removal rate, chipping, aspect ratio of micro-hole, machining, and forces involved in micro-cutting. Experimentation has been carried out with fabrication of simple microstructures such as micro-holes and micro-grooves using different construction material.

Microfeatures with better surface quality, geometrical shape, and lower chipping and tool wear have been achieved with vibration-assisted micro-drilling, micro-grinding, and micro-milling (Li and Wang 2014; Onikura et al. 2000, 2003; Lee et al. 2002). Elliptical vibrations have been used for ultrasonic vibration assisted micromachining process (Ammouri and Hamade 2012; Kim and Loh 2011, 2013; Brocato 2005; Brehl and Dow 2007). The out-of-phase excitations of the tool in the thrust force and cutting directions can result in a two-dimensional cutting trajectory with elliptical shape. The process is termed as two-dimensional vibration-assisted machining. A market improvement in surface accuracy over the conventional cutting process has been observed specially when employing worn tool. However, the generation of impulsive forces, because of the periodic contact between the tool and the workpiece, results in catastrophic fatigue failure of the tool. The reverse frictional force arising between the target material and the tool in elliptical vibration cutting results in decreased cutting resistance and improved surface finish.

Complex geometrical shapes could be machined using integrated arrangement comprising of ultrasonic vibration transducer and fast tool servo. The integrated arrangement has resulted in reduction in cutting force and, therefore, prolonged tool life (Chen et al. 2011). Significant degradation in machined shape of aluminum material machined using vibration-assisted micro-grooving has been revealed with larger two-dimensional error.

The ability of ultrasonic-assisted micro-lapping has been explored in producing array of micro-holes, rings, trenches, etc., on silicon. The size of microfeature ranged 50–500 μm and had aspect ratio ranging 1–5 (Zhang et al. 2006). Ultrasonic vibration assisted polishing tool was investigated for polishing of micro-molds. In this process, a three or five-axis work table is provided that mounts the small polishing tool. The tool is vibrated at ultrasonic frequency using piezoelectric actuators. The surface roughness of tungsten carbide was revealed to decrease with such vibration-assisted micromachining arrangement (Suzuki et al. 2006, 2010; Chee et al. 2011).

The machining performance of nontraditional micromachining process has been revealed to increase with the introduction of ultrasonic vibrations. As for instance, enhancement in machining performance with reduced machining time and increased stability has been achieved with vibration-assisted EDM process. There was neither any increase in surface roughness nor the tool electrode wear (Endo et al. 2008; Mahardika et al. 2012; Huang et al. 2003). A shorter machining time was reported for vibration-assisted micro-EDM process with perpendicular vibrations in comparison to the parallel vibrations (Endo et al. 2008). The machining rate associated with vibration-assisted micro-EDM of X210Cr12 was apparently increased with longer pulse time (25 μs). However, with pulse time greater than 25 μs, low accuracy in machining as well as surface quality has been reported (Ghiculescu et al. 2010). The dimensional integrity of machining tungsten carbide using vibration-assisted wire electrical discharge grinding (WEDG) was revealed to be increased and was far more superior to that machined without the use of ultrasonic vibrations (Hsue et al. 2012).

Vibration-assisted reverse micro-EDM was successfully employed for fabrication of micro-pillars with diameter ranging 40–50 μm. The desired surface texture for any cross-section micro-pillar was reported to be only achieved by the aforementioned vibration-assisted micromachining process (Mastud et al. 2012).

Surfaces with reduced surface roughness have been attained using the integrated vibration-assisted electrochemical micromachining (Ruszaj et al. 2003). However, the machining process must be carried out under suitable conditions as, for example, low amplitude of vibrations as well as medium frequency vibrations have been recommended for SS304. Such conditions aided in obtaining a more cylindrical micro-borehole having better surface finish with lesser overcut (Ghoshal and Bhattacharyya 2013, 2014).

## 5.2.2  Workpiece Vibration Assisted Micromachining

Vibrations to the workpiece material are provided by conversion of frequency oscillation associated with electric power to mechanical vibrations. The mechanical vibrations are obtained with the aid of boosters, horns, and piezoelectric transducers. Research in the domain of workpiece vibration assisted micromachining has been to achieve noncomplex shapes and to investigate the influence of vibrations on surface quality, tool life, micro-cutting forces, dimensional tolerances, aspect ratio, and material removal rate.

Workpiece elliptical vibration assisted micro-drilling and micro-milling have reported to machine AL6061-T6 with decreased surface roughness, tool life, and positioning error (Chern and Chang 2006; Chern and Lee 2006). Good surface roughness was achieved by optimizing amplitude of ultrasonic vibrations (Lian et al. 2013). Workpiece vibration assisted micro-drilling of glass resulted in reduced machining force and increased machining rate, depth, and tool life (Egashira et al. 2002). Workpiece vibration assisted micro-lapping has been employed successfully for production of micro-holes with very high precision. Micro-holes with improved surface roughness and roundness have been obtained using the aforementioned workpiece vibration assisted micromachining process (Wang et al. 2002).

Researchers have explored the potentiality and capability of workpiece vibration assisted micro-EDM process. For instance, eightfold machining efficiency for the workpiece vibration assisted micro-EDM process was achieved by Gao and Liu (2003). Significant improvements in material removal rate, surface quality, and dimensional accuracy was reported with deep drilling of micro-holes using workpiece vibration assisted micro-EDM process. Low-frequency vibrations were found to be more effective for workpiece vibration assisted micro-EDM process (Jahan et al. 2010a, b, 2011, 2012; Hung et al. 2006).

A comparative study of machining performance attained using workpiece vibration assisted micro-wire electrical discharge machining ($\mu$-WEDM) and tool vibration assisted $\mu$-WEDM for machining of Ti-6Al-4V was undertaken by Hoang and Yang (2013). A marked improvement in machining performance was reported in the former case, the machining performance in which case increased by 2.5 times compared to 1.5 times in the latter case.

The formation of surface oxides and recast layer was reported to be prevented in case of workpiece vibration assisted laser micromachining. Therefore, surface with improved quality was obtained by employing workpiece vibration assisted laser micromachining and it was higher than that achieved without the aid of ultrasonic vibrations (Kang et al. 2012; Zheng and Huang 2007).

### 5.2.3   Work Fluid Vibration Assisted Micromachining

Ultrasonic transducer attached to the tank containing work fluid is used to vibrate the contained work fluid inside the tank. Increased effectiveness in debris circulation has been reported in case of work fluid vibration assisted micro-EDM (Je et al. 2005; Kim et al. 2006). The investigation also revealed increased rate of material removal and higher geometrical accuracy. Vibrations in the work fluid could well be realized with the aid of horn (Ichikawa and Natsu 2013). The horn is provided with a small circular hole through which is fed the tool electrode. The ultrasonic vibrations can be applied to the work fluid without any contact between the vibrating horn and tool electrode. Such special arrangement has resulted in higher speed of machining, lesser tool wear rate, shorter lateral gap width, and smaller energy of discharge.

### 5.2.4   Objective Lens Vibration Assisted Micromachining

Park et al. (2012a, b) investigated the effectiveness of machining by employing vibrations to the objective lens. They proposed a novel micromachining process known as vibration-assisted femtosecond micro-drilling. Figure 5.1 delineates the process outline. The process was investigated for machining of copper substrate and it was revealed that features with improved surface finish and aspect ratio were obtained. Stirring effect emanating out of the vibrating objective lens enhances the heat transfer of the ablated particles and, therefore, results in improved surface quality and aspect ratio. The local particles are prevented from being joined to the wall surface as a result of increased cooling effect and, therefore, the surface of the wall appears to be cleaner than those obtained using laser micro-drilling without vibration of objective lens. The aspect ratio was reported to be increased by 160%.

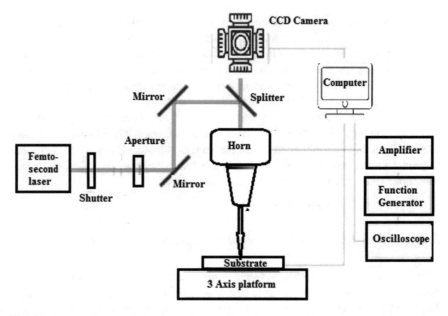

**Fig. 5.1** Process diagram of vibration-assisted variant of femtosecond laser micro-drilling machining process

## 5.3   External Electric Field Assisted Variant of Hybrid Micromachining

When electric field is introduced with the micromachining process, the debris formed during the machining process are removed efficiently and also the particles are prevented from being deposited back on the target surface. Application of external electric field to femtosecond laser for the material removal has been explored for silicon substrate (Zheng and Jiang 2009). Material removal through femtosecond laser is a complex phenomenon. The process talks place in a very short frame of time, i.e., at femtosecond time scale in which duration the free electrons absorb the laser energy and then in picosecond duration leave the workpiece surface. High peak power is produced as a result of ultrashort laser pulses possessing megajoules of pulse energy. Ionization of the parent material takes place as a result of high-intensity energy associated with the ultrashort laser. Formation of plasma occurs due to the ionization process. However, due to the presence of plasma a large number of particles can deposit themselves back to the machined surface. The process of redeposition can be inhibited by employing external electric field. As a result of electric field, external electric force acts on the particles. Only the particles with weight smaller than the electric force can be removed from the machining area. Particle weight can be calculated using Eq. (5.1) as follows:

**Fig. 5.2** Schematic of external electric field assisted variant of laser micromachining process

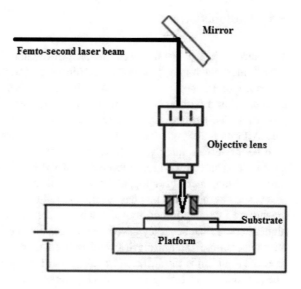

$$G = \left(\frac{r_p}{r_a}\right)^3 m_a g \tag{5.1}$$

where $r_a$ and $r_p$ are the radius of the Si atom and particle, respectively, $g$ is the acceleration due to gravity, and $m_a$ is the mass of the atom.

Surface with improved surface finish with reduced surface contaminants has been revealed to be obtained with the aforementioned hybrid micromachining process. Schematic diagram of external electric field assisted variant of laser micromachining is depicted in Fig. 5.2.

## 5.4 Carbon Nanofibre-Assisted Micromachining

Carbon nanofibers interlocks and arrange themselves in the form of micro-chains on application of electric field. As a result, a bridge is formed between the workpiece and the electrode. To reduce the insulating strength of the dielectric fluid in case of micro-EDM process, carbon nanofibers possessing excellent electrical conductivity have been employed. The introduction of carbon nanofibers also results in a decrease in spark gap between the workpiece and the electrode. Improved frequency of the discharge as well as the rate of material removal have been reported for RB-SiC machined using EDM process. Further, crater with reduced size on the surface of the workpiece have been observed owing to the multiple fine discharges. As a result, fine surface finish has been obtained with the process. Reduction in electrode wear as well as the concavity of the electrode tip has also been reported with the employability of nanofibers (Liew et al. 2013).

## 5.5   Conclusion

The present chapter has focussed on some of the major and least explored assisted hybrid micromachining processes. Machining performances have been reported to increase with the reported processes. Vibration-assisted micromachining processes have been used with various nontraditional and traditional cutting and machining processes. These processes have been categorized into tool vibration assisted, workpiece vibration assisted, and work fluid vibration assisted processes. Microfeatures with better surface quality, geometrical shape, and lower chipping and tool wear have been achieved with tool vibration assisted micromachining processes such as micro-drilling, micro-grinding, and micro-milling. Workpiece vibration assisted processes such as micro-drilling has resulted in reduced machining force and increased machining rate, depth, and tool life while increased rate of material removal and higher geometrical accuracy have been obtained using work fluid vibration assisted micromachining. External electric field assisted micromachining processes have been reported to machine the surface effectively with efficient flushing of debris particle. The introduction of carbon nanofibers in case of micro-EDM process has resulted in decrease in spark gap and hence enhanced rate of material removal. Efforts are still being made by the researchers to explore other assisted micromachine processes.

## References

S. Amini, M. Soleimani, H. Paktinat, M. Lotfi, Effect of longitudinal—torsional vibration in ultrasonic-assisted drilling. Mater. Manuf. Processes **32**(6), 616–622 (2017)

A.H. Ammouri, R.F. Hamade, BUEVA: a bi-directional ultrasonic elliptical vibration actuator for micromachining. Int. J. Adv. Manuf. Technol. **58**(9–12), 991–1001 (2012)

B. Azarhoushang, J. Akbari, Ultrasonic-assisted drilling of Inconel 738-LC. Int. J. Mach. Tools Manuf. **47**(7–8), 1027–1033 (2007)

D.E. Brehl, T.A. Dow, 3-D microstructure creation using elliptical vibration-assisted machining. ASPE Proc. Vibr. Assist. Mach. Technol. **21**, 26 (2007)

D.E. Brehl, T.A. Dow, Review of vibration-assisted machining. Precis. Eng. **32**(3), 153–172 (2008)

B.C. Brocato, *Micromachining using EVAM (Elliptical Vibration Assisted Machining)* (2005)

S.K. Chee, H. Suzuki, M. Okada, T. Yano, T. Higuchi, W.M. Lin, Precision polishing of micro mold by using piezoelectric actuator incorporated with mechanical amplitude magnified mechanism, in *Advanced Materials Research*, vol. 325 (Trans Tech Publications, 2011), pp. 470–475

H. Chen, M. Cheng, Y. Li, D. Zhang, Development of integrated precision vibration-assisted micro-engraving system. Trans. Tianjin Univ. **17**(4), 242 (2011)

W. Chen, X. Teng, L. Zheng, W. Xie, D. Huo, Burr reduction mechanism in vibration-assisted micro milling. Manuf. Lett. **16**, 6–9 (2018)

G.L. Chern, Y.C. Chang, Using two-dimensional vibration cutting for micro-milling. Int. J. Mach. Tools Manuf. **46**(6), 659–666 (2006)

G.L. Chern, H.J. Lee, Using workpiece vibration cutting for micro-drilling. Int. J. Adv. Manuf. Technol. **27**(7–8), 688–692 (2006)

K. Egashira, K. Mizutani, T. Nagao, Ultrasonic vibration drilling of microholes in glass. CIRP Ann. Manuf. Technol. **51**(1), 339–342 (2002)

T. Endo, T. Tsujimoto, K. Mitsui, Study of vibration-assisted micro-EDM—the effect of vibration on machining time and stability of discharge. Precis. Eng. **32**(4), 269–277 (2008)

C.R. Friedrich, P.J. Coane, M.J. Vasile, Micromilling development and applications for microfabrication. Microelectron. Eng. **35**(1–4), 367–372 (1997)

C. Gao, Z. Liu, A study of ultrasonically aided micro-electrical-discharge machining by the application of workpiece vibration. J. Mater. Process. Technol. **139**(1–3), 226–228 (2003)

D. Ghiculescu, N.I. Marinescu, S. Nanu, D. Ghiculescu, G. Kakarelidis, FEM study of synchronization between pulses and tool oscillations at ultrasonic aided microelectrodischarge machining. Rev. Tehnol. Neconventionale **14**(3), 19 (2010)

B. Ghoshal, B. Bhattacharyya, Influence of vibration on micro-tool fabrication by electrochemical machining. Int. J. Mach. Tools Manuf. **64**, 49–59 (2013)

B. Ghoshal, B. Bhattacharyya, Shape control in micro borehole generation by EMM with the assistance of vibration of tool. Precis. Eng. **38**(1), 127–137 (2014)

B. Ghoshal, B. Bhattacharyya, Vibration assisted electrochemical micromachining of high aspect ratio micro features. Precis. Eng. **42**, 231–241 (2015)

K.T. Hoang, S.H. Yang, A study on the effect of different vibration-assisted methods in micro-WEDM. J. Mater. Process. Technol. **213**(9), 1616–1622 (2013)

A.W.J. Hsue, J.J. Wang, C.H. Chang, Milling tool of micro-EDM by ultrasonic assisted multi-axial wire electrical discharge grinding processes. in *ASME 2012 International Manufacturing Science and Engineering Conference collocated with the 40th North American Manufacturing Research Conference and in participation with the International Conference on Tribology Materials and Processing* (American Society of Mechanical Engineers, 2012), pp. 473–479

H. Huang, H. Zhang, L. Zhou, H.Y. Zheng, Ultrasonic vibration assisted electro-discharge machining of microholes in Nitinol. J. Micromech. Microeng. **13**(5), 693 (2003)

J.C. Hung, J.K. Lin, B.H. Yan, H.S. Liu, P.H. Ho, Using a helical micro-tool in micro-EDM combined with ultrasonic vibration for micro-hole machining. J. Micromech. Microeng. **16**(12), 2705 (2006)

D. Huo, *Micro-cutting: Fundamentals and Applications* (Wiley, 2013)

T. Ichikawa, W. Natsu, Realization of micro-EDM under ultra-small discharge energy by applying ultrasonic vibration to machining fluid. Proc. CIRP **6**, 326–331 (2013)

M.P. Jahan, T. Saleh, M. Rahman, Y.S. Wong, Development, modeling, and experimental investigation of low frequency workpiece vibration-assisted micro-EDM of tungsten carbide. J. Manuf. Sci. Eng. **132**(5), 054503 (2010a)

M.P. Jahan, M. Rahman, Y.S. Wong, L. Fuhua, On-machine fabrication of high-aspect-ratio microelectrodes and application in vibration-assisted micro-electrodischarge drilling of tungsten carbide. Proc. Inst. Mech. Eng., Part B: J. Eng. Manuf. **224**(5), 795–814 (2010b)

M.P. Jahan, T. Saleh, M. Rahman, Y.S. Wong, Study of micro-EDM of tungsten carbide with workpiece vibration, in *Advanced Materials Research*, vol. 264, (Trans Tech Publications, 2011), pp. 1056–1061

M.P. Jahan, Y.S. Wong, M. Rahman, Evaluation of the effectiveness of low frequency workpiece vibration in deep-hole micro-EDM drilling of tungsten carbide. J. Manuf. Process. **14**(3), 343–359 (2012)

S.U. Je, H.S. Lee, C.N. Chu, D.W. Kim, Micro EDM with ultrasonic work fluid vibration for deep hole machining. J. Kor. Soc. Precis. Eng. **22**(7), 47–53 (2005)

B. Kang, G.W. Kim, M. Yang, S.H. Cho, J.K. Park, A study on the effect of ultrasonic vibration in nanosecond laser machining. Opt. Lasers Eng. **50**(12), 1817–1822 (2012)

G.D. Kim, B.G. Loh, Direct machining of micro patterns on nickel alloy and mold steel by vibration assisted cutting. Int. J. Precis. Eng. Manuf. **12**(4), 583–588 (2011)

G.D. Kim, B.G. Loh, Cutting force variation with respect to tilt angle of trajectory in elliptical vibration V-grooving. Int. J. Precis. Eng. Manuf. **14**(10), 1861–1864 (2013)

D.J. Kim, S.M. Yi, Y.S. Lee, C.N. Chu, Straight hole micro EDM with a cylindrical tool using a variable capacitance method accompanied by ultrasonic vibration. J. Micromech. Microeng. **16**(5), 1092 (2006)

T. Koyano, M. Kunieda, Ultra-short pulse ECM using electrostatic induction feeding method. Proc. CIRP **6**, 390–394 (2013)

B. Lauwers, Surface integrity in hybrid machining processes. Procedia Eng. **19**, 241–251 (2011)

J.S. Lee, D.W. Lee, Y.H. Jung, W.S. Chung, A study on micro-grooving characteristics of planar lightwave circuit and glass using ultrasonic vibration cutting. J. Mater. Process. Technol. **130**, 396–400 (2002)

K.M. Li, S.L. Wang, Effect of tool wear in ultrasonic vibration-assisted micro-milling. Proc. Inst. Mech. Eng., Part B: J. Eng. Manuf. **228**(6), 847–855 (2014)

C. Li, F. Zhang, B. Meng, L. Liu, X. Rao, Material removal mechanism and grinding force modelling of ultrasonic vibration assisted grinding for SiC ceramics. Ceram. Int. **43**(3), 2981–2993 (2017)

H. Lian, Z. Guo, Z. Huang, Y. Tang, J. Song, Experimental research of Al6061 on ultrasonic vibration assisted micro-milling. Proc. CIRP **6**, 561–564 (2013)

P.J. Liew, J. Yan, T. Kuriyagawa, Carbon nanofiber assisted micro electro discharge machining of reaction-bonded silicon carbide. J. Mater. Process. Technol. **213**(7), 1076–1087 (2013)

M. Mahardika, G.S. Prihandana, T. Endo, T. Tsujimoto, N. Matsumoto, B. Arifvianto, K. Mitsui, The parameters evaluation and optimization of polycrystalline diamond micro-electrodischarge machining assisted by electrode tool vibration. Int. J. Adv. Manuf. Technol. **60**(9–12), 985–993 (2012)

S. Mastud, M. Garg, R. Singh, J. Samuel, S. Joshi, Experimental characterization of vibration-assisted reverse micro electrical discharge machining (EDM) for surface texturing, in *ASME 2012 International Manufacturing Science and Engineering Conference collocated with the 40th North American Manufacturing Research Conference and in participation with the International Conference on Tribology Materials and Processing* (American Society of Mechanical Engineers), pp. 439–448

T. Moriwaki, E. Shamoto, Ultraprecision diamond turning of stainless steel by applying ultrasonic vibration. CIRP Ann. Manuf. Technol. **40**(1), 559–562 (1991)

M.D. Nguyen, Y. San Wong, M. Rahman, Profile error compensation in high precision 3D micro-EDM milling. Precis. Eng. **37**(2), 399–407 (2013)

H. Onikura, O. Ohnishi, Y. Take, A. Kobayashi, Fabrication of micro carbide tools by ultrasonic vibration grinding. CIRP Ann. Manuf. Technol. **49**(1), 257–260 (2000)

H. Onikura, R. Inoue, K. Okuno, O. Ohnishi, Fabrication of electroplated micro grinding wheels and manufacturing of microstructures with ultrasonic vibration, in *Key Engineering Materials*, vol. 238 (Trans Tech Publications, 2003), pp. 9–14

J.-K. Park, J.-W. Yoon, M.-C. Kang, S.-H. Cho, Surface effects of hybrid vibration-assisted femtosecond laser system for micro-hole drilling of copper substrate. Trans. Nonferrous Met. Soc. China **22**, s801–s807 (2012a)

J.K. Park, J.W. Yoon, S.H. Cho, Vibration assisted femtosecond laser machining on metal. Opt. Lasers Eng. **50**(6), 833–837 (2012b)

M. Rahman, A.S. Kumar, J.R.S. Prakash, Micro milling of pure copper. J. Mater. Process. Technol. **116**(1), 39–43 (2001)

P. Rodrigues, J.E. Labarga, Tool deflection model for micro milling process. Int. J. Adv. Manuf. Technol. **72**(5), 1–9 (2014)

A. Ruszaj, M. Zybura, R. Żurek, G. Skrabalak, Some aspects of the electrochemical machining process supported by electrode ultrasonic vibrations optimization. Proc. Inst. Mech. Eng., Part B: J. Eng. Manuf. **217**(10), 1365–1371 (2003)

E. Shamoto, N. Suzuki, T. Moriwaki, Y. Naoi, Development of ultrasonic elliptical vibration controller for elliptical vibration cutting. CIRP Ann. Manuf. Technol. **51**(1), 327–330 (2002)

H.S. Shin, M.S. Park, B.H. Kim, C.N. Chu, Recent researches in micro electrical machining. Int. J. Precis. Eng. Manuf. **12**(2), 371–380 (2011)

H. Suzuki, T. Moriwaki, T. Okino, Y. Ando, Development of ultrasonic vibration assisted polishing machine for micro aspheric die and mold. CIRP Ann. Manuf. Technol. **55**(1), 385–388 (2006)

H. Suzuki, T. Moriwaki, Y. Yamamoto, Y. Goto, Precision cutting of aspherical ceramic molds with micro PCD milling tool. CIRP Ann. Manuf. Technol. **56**(1), 131–134 (2007)

H. Suzuki, S. Hamada, T. Okino, M. Kondo, Y. Yamagata, T. Higuchi, Ultraprecision finishing of micro-aspheric surface by ultrasonic two-axis vibration assisted polishing. CIRP Ann. **59**(1), 347–350 (2010)

T. Tawakoli, B. Azarhoushang, M. Rabiey, Ultrasonic assisted dry grinding of 42CrMo4. Int. J. Adv. Manuf. Technol. **42**(9–10), 883–891 (2009)

E. Uhlmann, S. Piltz, K. Schauer, Micro milling of sintered tungsten–copper composite materials. J. Mater. Process. Technol. **167**(2–3), 402–407 (2005)

A.C. Wang, B.H. Yan, X.T. Li, F.Y. Huang, Use of micro ultrasonic vibration lapping to enhance the precision of microholes drilled by micro electro-discharge machining. Int. J. Mach. Tools Manuf. **42**(8), 915–923 (2002)

A. Weremczuk, R. Rusinek, J. Warminski, The concept of active elimination of vibrations in milling process. Proc. CIRP **31**, 82–87 (2015)

M. Xiao, K. Sato, S. Karube, T. Soutome, The effect of tool nose radius in ultrasonic vibration cutting of hard metal. Int. J. Mach. Tools Manuf. **43**(13), 1375–1382 (2003)

B. Xue, Y. Yan, J. Li, B. Yu, Z. Hu, X. Zhao, Q. Cai, Study on the micro-machining process with a micro three-sided pyramidal tip and the circular machining trajectory. J. Mater. Process. Technol. **217**, 122–130 (2015)

Y.B. Zeng, Q. Yu, S.H. Wang, D. Zhu, Enhancement of mass transport in micro wire electrochemical machining. CIRP Ann. Manuf. Technol. **61**(1), 195–198 (2012)

C. Zhang, E. Brinksmeier, R. Rentsch, Micro-USAL technique for the manufacture of high quality microstructures in brittle materials. Precis. Eng. **30**(4), 362–372 (2006)

H.Y. Zheng, H. Huang, Ultrasonic vibration-assisted femtosecond laser machining of microholes. J. Micromech. Microeng. **17**(8), N58 (2007)

H.Y. Zheng, Z.W. Jiang, Femtosecond laser micromachining of silicon with an external electric field. J. Micromech. Microeng. **20**(1), 017001 (2009)

M. Zhou, Y.T. Eow, B.K.A. Ngoi, E.N. Lim, Vibration-assisted precision machining of steel with PCD tools. Mater. Manuf. Processes **18**(5), 825–834 (2003)

# Chapter 6
# Combined Variant of Hybrid Micromachining Processes

## 6.1 Introduction

Continual advancements have been made in the variety of engineering domains demanding precision engineering. Engineering fields such as medical (Landolt et al. 2003), automobile (Li et al. 2006), and aviation (Okasha et al. 2010) are demanding for parts and features that are of micrometer range. As for instance micro-holes for fuel injection nozzles require high accuracy, sharp edges that are free from burrs and very small diameter usually ranging 0.4–0.8 mm. Further, higher surface finish with increased material removal rate is another major requirement of the industrial setup demanding micro-features and parts.

A number of techniques have been developed that meets the aforementioned industrial demands. As for instance, for production of precision small holes (Masuzawa 2000), technologies such as electrical discharge machining, laser machining, laser drill, etc. have been developed. However, such techniques have limited capabilities owing to the stringent technical requirements such as the requirement of sharp edges in micro-holes of the injection nozzle apart from the burr-free surface requirement. Often burr surfaces are obtained when using laser drilling or electrical discharge machining drilling. On the other hand, employability of electrochemical machining results in blunt edges. Also in some cases, the machining cost is much higher in comparison to their traditional counterparts. This cost may be attributed to the considerable wear and therefore breakage of the tools. Formation of the recast layer and heat-affected zones are some other disadvantages associated with the developed technologies.

To minimize the drawbacks associated with the individual processes and to reap the benefits of individual processes, combined variant of hybrid micromachining processes have been developed. These processes have the capability to machine hard-to-machine materials (Wang et al. 2003). Some of the examples of such processes include that of laser micro-drilling and jet electrochemical

© The Author(s), under exclusive license to Springer Nature Switzerland AG 2019
S. Bhowmik and D. Zindani, *Hybrid Micro-Machining Processes*,
SpringerBriefs in Applied Sciences and Technology,
https://doi.org/10.1007/978-3-030-13039-8_6

machining, micro-electrochemical machining, and micro-mechanical grinding and so on. Combined hybrid micromachining processes have not only enhanced the machining performance but have also satisfied the industrial demands of higher dimensional tolerances. Application of such processes has revealed that these processes are promising and robust.

Looking at the importance of combined hybrid micromachining processes, the present chapter outlines and discusses some of these processes. Applications of such combined arrangement have also been outlined towards the end of the chapter.

## 6.2   Laser Micro-drilling and Jet Electrochemical Machining

Two sources of energy from different sources are combined simultaneously in the laser micro-drilling and jet electrochemical machining. These energies are the energy of photons in laser drilling and energy of ions in electrochemical machining. Focussed beam of laser is aligned along the similar axis as that of the electrolyte jet that creates a contactless tool electrode. This coaxial arrangement strikes the same spot on the workpiece material. The laser beam results in heating and removal of material from the workpiece material and electrochemical reaction in the inter-pulse of laser is responsible for reduction in recast and spatter. The combined arrangement has been investigated for its capability by employing it on 321 stainless steel. It was reported that the spatter and recast were reduced by 93 and 90% respectively in comparison to the laser drilling being carried out under atmospheric conditions (Zhang et al. 2009; Hua and Jiawen 2010; Zhang and Xu 2012). The process outline is depicted in Fig. 6.1.

## 6.3   Micro-electrochemical Machining Combined with Micro-mechanical Grinding

A combined variant of hybrid micromachining arrangement of material removal constituting electrochemical machining and micro-mechanical grinding was established in the year 2011. The process setup is outlined in Fig. 6.2. In such a combined arrangement, diamond abrasives are coated on a spherical metal rod. The higher speed rotation of cathodic tool results in material removal that takes place mechanically and electrochemically. The core of the tool is required to be conductive whereas the nonconductivity of abrasive grains should be ensured in carrying out the combined process effectively. As for instance, the spherical metal rod is coated with nickel and the abrasive particles can be diamond particles. Workpiece material acts as an anode whereas the tool is made cathode and is charged negatively. The nonconductive abrasive particles project beyond the conductive bond surface. Therefore while machining

**Fig. 6.1**  Process outline of laser micro-drilling and jet electrochemical machining

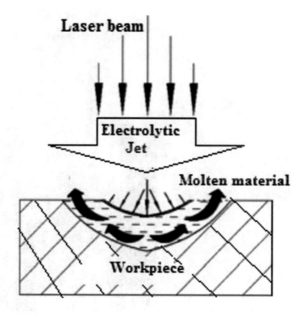

a hole, a small gap is established between the tool nickel layer and the side wall of the hole. The material removal process takes place in two phases: electrochemical action in phase 1 and combination of mechanical grinding and electrochemical reaction in phase 2. When the gap between the hole and tool is filled with electrolytic solution and the tool is charged, the electrochemical action begins. Formation of passive film takes place on the surface of the hole. In phase 2 the mechanical grinding process also gets involved with the electrochemical action in the removal of the material from the workpiece surface. The gap decrease between the tool and the workpiece surface as the abrasive tool descends the hole. Decent continues until the bottom of the abrasive tool comes in contact of the workpiece surface. As a result of this contact, the nonreactive passive layer of the hole surface gets removed owing to the grinding action of the abrasive grains. Phase 2 terminates where the maximum tool diameter is reached. The tool is required to be insulated in order to achieve high dimensional accuracy and sharp edges. There will be another phase, i.e., phase 3 in case the tool is not insulated and in this phase no material removal will take place. However, due to the presence of electric field between surface of machined hole and the cathodic tool the electrochemical dissolution will continue resulting in formation of taper hole. Surface finish of up to 0.21 µm has been reported to achieve in stainless steel 321 workpiece material (Zhu et al. 2011).

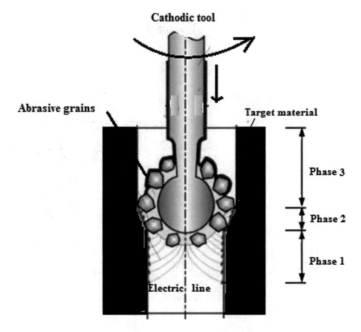

**Fig. 6.2** Process setup of micro-electrochemical machining combined micro-mechanical grinding

## 6.4    Micro-electrochemical Discharge Machining

This combined hybrid micromachining arrangement is made from combined electrochemical reaction and electrodischarge action. Ionic gas bubbles charged positively such as hydrogen is generated owing to the electrochemical action. As the DC voltage is applied between the cathodic tool and anodic workpiece material, breakdown of the gas bubble insulating layer takes place resulting in electrical discharge between the tool and the workpiece material. As a result of the electrical discharge, melting and vaporization of the workpiece material take place resulting in the material removal process.

A number of connotations have been used for this combined variant of hybrid micromachining processes such as spark-assisted chemical graving, electrochemical arc machining, and electrochemical spark machining (Wüthrich and Fascio 2005). Different aspects related to micro-electrochemical discharge machining have been investigated by past researchers. These include film thickness of gas (Wüthrich and Hof 2006; Zheng et al. 2008; Cheng et al. 2010a, b), wettability of electrodic tool (Wüthrich et al. 2005), various process parameters, and their characterisation (Sarkar et al. 2006; Maillard et al. 2007; Manna and Narang 2012; Ziki and Wüthrich 2013; Paul and Hiremath 2013; Razfar et al. 2013), mechanism of material removal (Jalali et al. 2009), application of rectangular voltage pulses, and its impact on heat-affected zone (Kim et al. 2006), transition voltage (Cheng et al. 2010a, b), use of resistor–capacitor circuit (Sarkar et al. 2009), heat transfer simulation (Krötz et al. 2013), and applicability of the process for different materials with different shapes (West and Jadhav 2007; Furutani and Maeda 2008; Zheng et al. 2007; Cao et al. 2009; Coteaţă et al. 2011; Kulkarni et al. 2011; Ziki et al. 2012; Huang et al. 2014; Lijo and Hiremath 2014). The studies have revealed that material removal rate, thickness of heat-affected zone and radial cut are significantly affected by applied voltage in comparison to other machining process parameters such as concentration of electrolytic solution, inter-electrode gap and depth of tool immersion (Sarkar et al. 2006; Razfar et al. 2013).

## 6.5  Simultaneous Micro-electrical Discharge Machining and Micro-electrochemical Machining

The material removal process in the combined hybrid micromachining setup takes place in deionised water possessing low resistivity. In this combined arrangement the electrochemical reaction takes place in tandem with the electrical discharge process. The original shape in this arrangement is maintained since the material removal process takes place layer-by-layer (Nguyen et al. 2012a, b, 2013a, b).

The simultaneous arrangement has been successfully implemented for the production of micro-shapes that possesses improved dimensional accuracy and surface integrity. On implementing the process for SUS304, surface roughness of the order of 22 nm have been reported to achieve. The simultaneous process, however, suffers from certain drawbacks such as damage to the workpiece surface and excessive erosion caused by the electrolytic solution. These drawbacks may be attributed to the slight electrical conductivity in deionised water. Excessive erosion has however been suppressed with application of nanosecond voltage pulse. Usage of epoxy resin side insulated electrodic tool has further minimized the problem of electrolytic erosion (Yin et al. 2014).

## 6.6  Micro-electrical Discharge Machining Combined with Electrorheological Fluid-Assisted Polishing

The possibility of inducing electric field to aid polishing in micro-electrical discharge machining (micro-EDM) process has been explored by the scientific community giving birth to micro-EDM and electrorheological fluid-assisted polishing process. In this combined arrangement, anode is formed by the micro-tool electrode (copper) and the conductive workpiece material is made the cathode. A narrow gap is thus formed between the two electrodes. The material is liquefied and immediately vaporized as a result of very intense heat energy produced in the gap owing to the applied external electric field. Clustering of the dispersed particle takes place in the form of fibrous

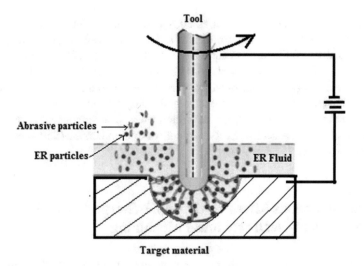

**Fig. 6.3** Micro-EDM combined with electrorheological fluid-assisted polishing process

structure and they align themselves along the electric line. The fibrous structures align themselves in line with the abrasive particles present in the ER fluid. Abrasive particles indent the workpiece surface when the chain of particles are made to come in contact with the workpiece surface through the means of rotating micro-tool. Alumina and SiC have been explored and used as abrasive particles in this combined micromachining technique (Tsai et al. 2008; Zhao et al. 2012). Figure 6.3 depicts the micro-EDM combined with electrorheological fluid-assisted polishing process.

## 6.7   Applications

Advantages of standalone micromachining processes and other assisted micromachining processes can be explored on integration with the various hybrid micromachining process. The integrated framework tends to enhance the benefits and reduce the disadvantages associated with individual hybrid techniques. Newer materials with enhanced mechanical properties and of various shapes can be machined with higher precision and enhanced surface integrity and quality using such an integrated arrangement. Following discussion outlines few research work that has been carried out in combined hybrid micromachining process with the other standalone micromachining and assisted processes.

Performance of various processes has been reported to increase with the application of magnetic field and vibrations. As for instance, the machining depth in case of micro-EDM process has been improved by integrating it with the ultrasonic-vibration electrolyte. Ultrasonic-vibration of electrolytic solution ensures its flow adequacy for generation of the spark and efficient removal of chips from the tool and workpiece gap (Han et al. 2009). Enhanced electrolytic circulation has also been achieved with the application of magnetic field that is induced through the magneto-hydrodynamic convection. The enhance circulation of electrolytic solution results in enhanced machining performance. An improvement of nearly 24% in geometry accuracy has been achieved with the magnetic field assisted micro-EDM. Nearly 55% improvement in machining time was also reported in this case (Cheng et al. 2010a, b). Increase in machining depth ranging 300–520 μm has been reported in vibration-assisted micro-ECDM drilling of borosilicate glass (Han et al. 2009).

Another integrated arrangement investigated by researchers is that where electro-chemical discharge milling process has been employed for machining of microstruc-tures and then by PCD tools. The process has been applied on glass and improved surface roughness of 0.05 μm has been achieved (Cao et al. 2013). Micro-holes with the high quality of surface have been achieved by combining wire electrical discharge grinding and electrical discharge machining with high-frequency dither grinding. The integrated framework consists of electrical discharge machine setup, a four-axis control system, wire electrical discharge grinding mechanism, and mechanism of high-frequency dither grinding. Dither mechanism is composed of power ampli-fier, piezoelectric ceramics, function generator, and reeds. Wire electrical discharge grinding is fixed on electrical discharge machining worktable. Electrical discharge machining head carries the four-axis control system.

Wire electrical discharge grinding is used to machine a circular rod of tungsten carbide. Micro-holes are then drilled using the machine circular rod. Slurry in the tank carries alumina particles. Polishing of micro-holes is carried out using the coupled electrical discharge machining and high-frequency dither grinding mechanism. The eccentricity problem associated with the rotating tool is eliminated as the integrated framework does not involve dismounting of the electrode or tool. Introduction of high-frequency dither grinding has resulted in decreased surface roughness from 2.12 to 0.85 μm for Hymu 80 workpiece (Liu et al. 2006).

A number of problems have been revealed with the generation of cooling holes having diameter less than 1 mm in nickel alloy coated with nonconductive ceramics. Since the ceramic material is alumina electrical discharge machining is not viable because of electrically insulating nature of the alumina particles. The specific elec-trical resistance of alumina is greater than 100 Ω cm. The associated problem can be reduced with the employability of ultrasonic machining to penetrate the alumina coating and then utilization of tool-vibration-assisted micro-EDM drilling. However, to achieve effective USM/EDM machining performance, usage of appropriate tool material is critical. Mild steel has been proven as one of the effective tool material for the integrated machining process (Thoe et al. 1999).

## 6.8 Conclusion

This chapter has illuminated on least explored combined hybrid micromachining processes wherein the total material removal rate is the sum total of the material removal achieved using individual micromachining processes. Therefore one of the associated advantages of the combined arrangement is the enhanced material removal rate. The arrangement minimizes the disadvantages associated with the individual processes, therefore, the overall machining performance is enhanced. Improved surface finish is another major advantage associated with such combined micromachining arrangement. Effectiveness of debris flushing is improved. However, such processes suffer from certain disadvantages such as complex machining mechanism, high setup cost, and maintenance intensive. A number of applications for combined and assisted hybrid processes, sequential and assisted hybrid processes, and combined and sequential hybrid processes have also been discussed towards the end of the chapter.

# References

X.D. Cao, B.H. Kim, C.N. Chu, Micro-structuring of glass with features less than 100 μm by electrochemical discharge machining. Precis. Eng. **33**(4), 459–465 (2009)

X.D. Cao, B.H. Kim, C.N. Chu, Hybrid micromachining of glass using ECDM and micro grinding. Int. J. Precis. Eng. Manuf. **14**(1), 5–10 (2013)

C.P. Cheng, K.L. Wu, C.C. Mai, C.K. Yang, Y.S. Hsu, B.H. Yan, Study of gas film quality in electrochemical discharge machining. Int. J. Mach. Tools Manuf. **50**(8), 689–697 (2010a)

C.P. Cheng, K.L. Wu, C.C. Mai, Y.S. Hsu, B.H. Yan, Magnetic field-assisted electrochemical discharge machining. J. Micromech. Microeng. **20**(7), 075019 (2010b)

M. Coteață, H.P. Schulze, L. Slătineanu, Drilling of difficult-to-cut steel by electrochemical discharge machining. Mater. Manuf. Processes **26**(12), 1466–1472 (2011)

K. Furutani, H. Maeda, Machining a glass rod with a lathe-type electro-chemical discharge machine. J. Micromech. Microeng. **18**(6), 065006 (2008)

M.S. Han, B.K. Min, S.J. Lee, Geometric improvement of electrochemical discharge micro-drilling using an ultrasonic-vibrated electrolyte. J. Micromech. Microeng. **19**(6), 065004 (2009)

Z. Hua, X. Jiawen, Modeling and experimental investigation of laser drilling with jet electrochemical machining. Chin. J. Aeronaut. **23**(4), 454–460 (2010)

S.F. Huang, Y. Liu, J. Li, H.X. Hu, L.Y. Sun, Electrochemical discharge machining micro-hole in stainless steel with tool electrode high-speed rotating. Mater. Manuf. Processes **29**(5), 634–637 (2014)

M. Jalali, P. Maillard, R. Wüthrich, Toward a better understanding of glass gravity-feed micro-hole drilling with electrochemical discharges. J. Micromech. Microeng. **19**(4), 045001 (2009)

D.J. Kim, Y. Ahn, S.H. Lee, Y.K. Kim, Voltage pulse frequency and duty ratio effects in an electrochemical discharge microdrilling process of Pyrex glass. Int. J. Mach. Tools Manuf **46**(10), 1064–1067 (2006)

H. Krötz, R. Roth, K. Wegener, Experimental investigation and simulation of heat flux into metallic surfaces due to single discharges in micro-electrochemical arc machining (micro-ECAM). Int. J. Adv. Manuf. Technol. **68**(5–8), 1267–1275 (2013)

A.V. Kulkarni, V.K. Jain, K.A. Misra, Electrochemical spark micromachining (microchannels and microholes) of metals and non-metals. Int. J. Manuf. Technol. Manage. **22**(2), 107–123 (2011)

D. Landolt, P.F. Chauvy, O. Zinger, Electrochemical micromachining, polishing and surface structuring of metals: fundamental aspects and new developments. Electrochim. Acta **48**(20–22), 3185–3201 (2003)

L. Li, C. Diver, J. Atkinson, R. Giedl-Wagner, H.J. Helml, Sequential laser and EDM micro-drilling for next generation fuel injection nozzle manufacture. CIRP Ann. Manuf. Technol. **55**(1), 179–182 (2006)

P. Lijo, S.S. Hiremath, Characterisation of micro channels in electrochemical discharge machining process, in *Applied Mechanics and Materials*, vol. 490 (Trans Tech Publications, 2014), pp. 238–242

H.S. Liu, B.H. Yan, C.L. Chen, F.Y. Huang, Application of micro-EDM combined with high-frequency dither grinding to micro-hole machining. Int. J. Mach. Tools Manuf. **46**(1), 80–87 (2006)

P. Maillard, B. Despont, H. Bleuler, R. Wüthrich, Geometrical characterization of micro-holes drilled in glass by gravity-feed with spark assisted chemical engraving (SACE). J. Micromech. Microeng. **17**(7), 1343 (2007)

A. Manna, V. Narang, A study on micro machining of e-glass–fibre–epoxy composite by ECSM process. Int. J. Adv. Manuf. Technol. **61**(9–12), 1191–1197 (2012)

T. Masuzawa, State of the art of micromachining. CIRP Ann. Manuf. Technol. **49**(2), 473–488 (2000)

M.D. Nguyen, M. Rahman, Y. San Wong, Enhanced surface integrity and dimensional accuracy by simultaneous micro-ED/EC milling. CIRP Ann. Manuf. Technol. **61**(1), 191–194 (2012a)

M.D. Nguyen, M. Rahman, Y. San Wong, Simultaneous micro-EDM and micro-ECM in low-resistivity deionized water. Int. J. Mach. Tools Manuf. **54**, 55–65 (2012b)

M.D. Nguyen, M. Rahman, Y. San Wong, Modeling of radial gap formed by material dissolution in simultaneous micro-EDM and micro-ECM drilling using deionized water. Int. J. Mach. Tools Manuf. **66**, 95–101 (2013a)

M.D. Nguyen, M. Rahman, Y. San Wong, Transitions of micro-EDM/SEDCM/micro-ECM milling in low-resistivity deionized water. Int. J. Mach. Tools Manuf. **69**, 48–56 (2013b)

M.M. Okasha, P.T. Mativenga, N. Driver, L. Li, Sequential laser and mechanical micro-drilling of Ni superalloy for aerospace application. CIRP Ann. **59**(1), 199–202 (2010)

L. Paul, S.S. Hiremath, Response surface modelling of micro holes in electrochemical discharge machining process. Proc. Eng. **64**, 1395–1404 (2013)

M.R. Razfar, J. Ni, A. Behroozfar, S. Lan, An investigation on electrochemical discharge micro-drilling of glass, in *ASME 2013 International Manufacturing Science and Engineering Conference collocated with the 41st North American Manufacturing Research Conference* (American Society of Mechanical Engineers, 2013), p. V002T03A013

B.R. Sarkar, B. Doloi, B. Bhattacharyya, Parametric analysis on electrochemical discharge machining of silicon nitride ceramics. Int. J. Adv. Manuf. Technol. **28**(9–10), 873–881 (2006)

B.R. Sarkar, B. Doloi, B. Bhattacharyya, Investigation into the influences of the power circuit on the micro-electrochemical discharge machining process. Proc. Inst. Mech. Eng., Part B: J. Eng. Manuf. **223**(2), 133–144 (2009)

T.B. Thoe, D.K. Aspinwall, N. Killey, Combined ultrasonic and electrical discharge machining of ceramic coated nickel alloy. J. Mater. Process. Technol. **92**, 323–328 (1999)

Y.Y. Tsai, C.H. Tseng, C.K. Chang, Development of a combined machining method using electrorheological fluids for EDM. J. Mater. Process. Technol. **201**(1–3), 565–569 (2008)

Z.Y. Wang, K.P. Rajurkar, J. Fan, S. Lei, Y.C. Shin, G. Petrescu, Hybrid machining of Inconel 718. Int. J. Mach. Tools Manuf. **43**(13), 1391–1396 (2003)

J. West, A. Jadhav, ECDM methods for fluidic interfacing through thin glass substrates and the formation of spherical microcavities. J. Micromech. Microeng. **17**(2), 403 (2007)

R. Wüthrich, V. Fascio, Machining of non-conducting materials using electrochemical discharge phenomenon—an overview. Int. J. Mach. Tools Manuf. **45**(9), 1095–1108 (2005)

R. Wüthrich, L.A. Hof, The gas film in spark assisted chemical engraving (SACE)—a key element for micro-machining applications. Int. J. Mach. Tools Manuf. **46**(7–8), 828–835 (2006)

R. Wüthrich, L.A. Hof, A. Lal, K. Fujisaki, H. Bleuler, P. Mandin, G. Picard, Physical principles and miniaturization of spark assisted chemical engraving (SACE). J. Micromech. Microeng. **15**(10), S268 (2005)

Q. Yin, B. Wang, Y. Zhang, F. Ji, G. Liu, Research of lower tool electrode wear in simultaneous EDM and ECM. J. Mater. Process. Technol. **214**(8), 1759–1768 (2014)

H. Zhang, J. Xu, Laser drilling assisted with jet electrochemical machining for the minimization of recast and spatter. Int. J. Adv. Manuf. Technol. **62**(9–12), 1055–1062 (2012)

H. Zhang, J. Xu, J. Wang, Investigation of a novel hybrid process of laser drilling assisted with jet electrochemical machining. Opt. Lasers Eng. **47**(11), 1242–1249 (2009)

Y.W. Zhao, D.X. Geng, X.M. Liu, Study on combined process of micro-EDM and electrorheological fluid-assisted polishing. In *Advanced Materials Research*, vol. 418 (Trans Tech Publications, 2012), pp. 1167–1170

Z.P. Zheng, W.H. Cheng, F.Y. Huang, B.H. Yan, 3D microstructuring of Pyrex glass using the electrochemical discharge machining process. J. Micromech. Microeng. **17**(5), 960 (2007)

Z.P. Zheng, J.K. Lin, F.Y. Huang, B.H. Yan, Improving the machining efficiency in electrochemical discharge machining (ECDM) microhole drilling by offset pulse voltage. J. Micromech. Microeng. **18**(2), 025014 (2008)

D. Zhu, Y.B. Zeng, Z.Y. Xu, X.Y. Zhang, Precision machining of small holes by the hybrid process of electrochemical removal and grinding. CIRP Ann. Manuf. Technol. **60**(1), 247–250 (2011)

J.D.A. Ziki, R. Wüthrich, Forces exerted on the tool-electrode during constant-feed glass microdrilling by spark assisted chemical engraving. Int. J. Mach. Tools Manuf. **73**, 47–54 (2013)

J.D.A. Ziki, T.F. Didar, R. Wüthrich, Micro-texturing channel surfaces on glass with spark assisted chemical engraving. Int. J. Mach. Tools Manuf. **57**, 66–72 (2012)

Printed in the United States
By Bookmasters